Darwin Meets Einstein

Darwin Meets Einstein

On the Meaning of Science

Frans W. Saris

AMSTERDAM UNIVERSITY PRESS

English translation: Pien Saris-Bertelsmann

Cover design: NEON, ontwerp en communicatie, Amsterdam
Layout: Het Steen Typografie, Maarssen

ISBN 978 90 8964 058 1
e-ISBN 978 90 4850 830 3
NUR 911

*To colleagues and friends, for decades of Fun,
Utilization, Theories of everything and Survival,
Thank you*

Contents

Theories of Everything

Survival

The temple of science

In a famous speech to the German Physical Society in 1918, Albert Einstein honoured his colleague Max Planck with the following words:

'In the temple of science are many mansions, and various indeed are they that dwell therein and the motives that have led them thither. Many take to science out of a joyful sense of superior intellectual power; science is their own special sport to which they look for vivid experience and the satisfaction of ambition; many others are to be found in the temple who have offered the products of their brains on this altar for purely utilitarian purposes. Were an angel of the Lord to come and drive all the people belonging to these two categories out of the temple, the assemblage would be seriously depleted, but there would still be some men, of both present and past times, left inside. Our Planck is one of them, and that is why we love him.'

So, Einstein distinguished three motives. Scientists do research either for Fun, or for Utilization, or for what motivated Planck and Einstein most: a Theory of Everything. Because he was Einstein, many scientists and especially physicists followed him in this preference. The first motive, fun, also stands for fundamental research, science for science's sake, which in the hierarchy of academia stands well above applied research. Yet, in the temple of science the search for the Theory of Everything is still of saintly stature, despite the fact that after almost a century this theory still does not exist.

As an experimental physicist I have also offered my products on the altar. They consisted not only of purely scientific and technical publications and patents; I have felt the need to share my work with a general public in the form of essays and columns in newspapers, literary journals and books. Reflecting on my work I recognized the three motives of Albert Einstein, but more recently I have discovered a fourth.

Reading and thinking about the history, philosophy and sociology of science I have learned to ask four questions. The first is: What is the immediate reason for humans to carry out scientific research? That is very different for different people. There are those who do it out of competitiveness: they want to be first in their field, and for them it is only the first prize that counts. Others are more like pioneers; they want to be first to reach a peak or an uninhabited area. Yet others are much more motivated to improve our world, they practice science for the benefit of mankind.

My second question is: How does inquiry, the behaviour of scientists, come about? Is it taught or inherited? Both surely. People are born as curious animals, and for some of them science is a passion, so that in principle education should stimulate inquiry. Curiosity is important, if you don't look you won't find anything at all, but if on the other hand you *are* looking for something, you will frequently find something very different from what you had expected. Once you experience the thrill of the creative moment, you become intoxicated, even hooked.

My third question: How did scientists *develop*? How did the practice of science evolve over the course of time? In the early days scientists still worked individually, today they perform their work in orchestrated groups from very different laboratories all over the world, all of whom try to solve the same given problem collaboratively. 'Big Science for Big Business' is not only true in physics. Even life scientists have to work in this way and in due course the social sciences and even the humanities will have to follow suit.

My fourth and for me by far most important question is this: What is the function of science? Today scientists have lost track of what they are supposed to be doing. Science has become a contest, a hype, and as a result a certain decadence has set in. No wonder society at large does not value scientific research and education as much as it should. In Darwinian thinking science contributes to our survival. That is its function. Individual behaviour in our culture displays many mutations. Most mutations are selected out: they melt like snow in the sun, except for those behaviours, and that culture, which truly contribute to our survival. That culture will certainly survive. This holds especially true for the sciences. In our post-modern times we have lost the meaning of science, because science is not for competition, not for creating wealth, not even for fun, but science is for survival, the survival of humans, the survival of life on earth.

Darwin Meets Einstein is a collection of essays, columns and a play *On the Meaning of Science*, which I have ordered according to Einstein's three motives, Fun, Utilization and Theory of Everything, and inspired by Darwin I have added a fourth section on science for Survival.

Frans W. Saris
August 2009

*To colleagues and friends, for decades of **Fun**,*
Utilization, Theories of everything and Survival,
Thank you

Diary of a physicist

Moscow, 19 September 1977. Rumours are buzzing all over. There are Russians, an East German and a Hungarian with contributions on laser-annealing of Silicon. This could mean a revolution for the semiconductor industry, but all papers are presented in Russian so nobody can follow them and the lobby is busier than the conference-room. Apparently a certain Chaibullin from Kazan discovered already in 1974 that defects in crystalline silicon may be annealed with a short light pulse from a ruby laser. The Americans read the Russian journals just as poorly as I do, for Chaibullin's discovery was not noted until Rimini, from the Catania group, talked about it in a colloquium last month in the group of Walter Brown at Bell Labs. The story goes that halfway through his colloquium Rimini only had the front rows of his audience still present. The Bell boys in the back had sneaked out of the room to quickly do an experiment with a laser and submit an article to Applied Physics Letters before Rimini's colloquium was finished. It means that we should not get into this, for much has been done already and too many groups will compete in laser-annealing.

Moscow, 22 September 1977. I was present at an interesting but laborious discussion between Chaibullin, Rimini and Brown on the physical mechanism of laser-annealing. Rimini and Brown think that the phenomena Chaibullin has measured are easily explained by assuming that under the laser pulse silicon locally gets so hot that it melts and a liquid silicon layer is formed on top of the crystal. After the laser pulse stops the liquid layer solidifies and takes on the same structure as the underlying crystal. Chaibullin says 'njet' many times and with some emphasis. His English, however, is not good enough to really argue with the people around him. Walter Brown sticks his pen back into his shirt pocket, the well-known sign that for him the discussion is finished and says: 'Well, that leaves something for us to prove.' Or for us, I think!

Amsterdam, 3 October 1977. The mail in the lab this morning made me quickly forget that fantastic trip to Moscow, Leningrad and Uzbekistan.

Philips announced that they would terminate our contract on ion implantation research. My budget is fully dependent on them, and not only my budget. I have immediately asked Dick Hoonhout to start work on laser annealing for his PhD thesis. Not just because here is an interesting subject not yet completely run down, but also because Philips will certainly be interested in this, as laboratories of IBM, Bell, Cornell and in Stanford have begun also.

Amsterdam, 16 October 1977. Yesterday, Dick Hoonhout, together with Yde Tamminga of the Philips group did the first experiments with a ruby laser to re-crystallize a silicon surface layer amorphized by ion implantation. Colour changes of the silicon surface clearly indicated that it worked. Today, Dick suggested to also try it on amorphous deposited layers. A great idea, why didn't I think of it?

Amsterdam, 21 November 1977. Great News! While I was muttering to Dick about the deposition equipment he says: 'But I can simply take the sample we still have in the drawer, on which we had deposited an amorphous layer last month.' To give it a try Dick fired two laser pulses on this piece of silicon, and yes, we see the same colour changes as on the implanted samples. Colour changes in Dick's face prove that he has experienced that fantastic flash of enthusiasm which belongs to a creative moment in research.

In spite of all discussions in the philosophy of science, in spite of the student revolution in the 1960s and in spite of the opposition from politics and society at large, the traditional view of science by scientists themselves has not changed and is propagated in the journals and meetings of learned societies, via annual reports and other propaganda from our research organizations and in ceremonial speeches by professors. I cannot and will not take part in this, for this is not the view I have of my own activity as a scientist and because I believe that disseminating the traditional view of science is harmful.

Perhaps it is unusual on the occasion of an inaugural lecture, but I will try to contribute to a more realistic picture of the job of the physicist by reading a few passages from my diary. These concern our research on laser-annealing, which exhibits every aspect of modern science. Most respected universities in the world and most research labs of the big electronic industries and research organizations work in this field. The productivity is enormous: more than three hundred scientific publications and dozens of patents every year. The contribution from my own research lab is some fifteen scientific papers, one PhD thesis and a new method to fabricate silicon solar cells. Here, however, I do not care so much about the results, rather the way they were obtained.

The thinking behind it. Therefore I have chosen especially those parts of my diary that give an impression of the human aspects that play a significant role in scientific research, such as rationality and belief, motivation and frustration, innovation and conservatism, co-operation and competition.

I believe that in these parts of my diary few indications may be found for the popular view of the values-free scientist, who, motivated purely by curiosity and guided only by rationality, comes to a scientific decision on the choice of his field of research.

Let us now look how this scientist, who has just made a discovery, is conscious of the possible applications of his research, to the extent that the well-known utilization process can start leading to a healthy interaction between the so-called 'technology push and market pull' through which transfer of valuable scientific information towards industry takes place.

Amsterdam, 21 November 1977. [...] We will not be able to reproduce these fantastic results in the near future. We don't have any more samples in the drawer and to repair our deposition system will take months. We cannot wait so long with claiming laser-annealing of deposited layers, can we? Therefore, I went to Hofker to suggest that through Philips he will file a patent of our invention, and he agreed.

Amsterdam, 22 November 1977. Reading 'News and Discoveries' in Physics Today, I had the idea of sending out a press release on 'Crystal growth of silicon with pulsed laser radiation'. Because in our country American methods are really considered as bragging, I thought it wiser not to submit this to the Dutch Journal of Physics by myself but to ask Jan Heyn of the FOM Organization to distribute the press release under his own name.

Amsterdam, 1 December 1977. Last week I saw an American paper with a sketch of a fully automated production line for silicon solar cells, fabricated by ion implantation and electron-beam annealing. Today I have submitted a research proposal to the European Community in Brussels with a sketch of just such a production line but based on laser-annealing of deposited layers. This should guarantee the extension of our EU contract. For that matter, a variation of this proposal could also be submitted to FOM for its new programme on Technical Physics and Innovation.

Amsterdam, 12 January 1978. Last week I thought: Let us hope that damned deposition system starts working again soon, then Dick can reproduce his results and we may finally submit a paper. Yesterday, at home I read in

Applied Physics Letters an article from the group at Hughes in which the authors show that amorphously deposited films may be crystallized through heating. These people only need to get the idea of using a laser and we will be lost! Immediately I called Dick Hoonhout at home and asked him to start writing a draft of a letter for which I promised to write the introduction. This morning, to my surprise, Dick put the draft on my desk, so that I will have to be fast producing the intro.

Amsterdam, 19 January 1978. Today we have submitted our paper to professor Volger, deputy director of Philips Research and editor of Physics Letters. In this way, I think we kill two birds with one stone: our paper will appear within a few weeks, thus we will stay ahead of the American competition, and secondly, these important results for Philips immediately are put on the table of the Philips directorate. In the introduction I mention ultra-fast transistors, solar cells and silicon superlattices as possible applications. They should be duly impressed!

Amsterdam, 26 January 1978. How the hell is this possible! Yesterday, our research proposal to the European Community was rejected. This morning's mail brought a copy of the patent application with a note that it would only be considered after electrical measurements. This afternoon, the Philips deputy director who is responsible for our contract suddenly visited us. Without showing any reaction he listened to the enthusiastic story of Dick Hoonhout. After Dick had left the room he says: 'You are wasting your time, for fifteen years we have worked on films of Silicon and it never came to anything useful. Moreover, if silicon melts under the laser pulse, this is about the worst you can do to a wafer.'

Amsterdam, 12 February 1978. I am called to Jaap Kistemakers' office. There lies our article for Physics Letters, rejected! The reason is that the 'exorbitant expectations of the introduction are not supported by the rest of the paper'. Not knowing what to do and totally depressed I tell Jaap of my European contract, the patent application and the visit from Philips Research. I am finished. One month ago appointed to deputy director especially for contract research and already the only two contracts we have are practically gone. Jaap cheers me up and says: 'Leave that little paper with me and you go to Philips in Paris where the experts on solar cells are, good luck.'

Amsterdam, 22 February 1978. I am furious. Jaap has completely rewritten the introduction to our paper and resubmitted this to Volger, who accepted it this time.

I have also been to Paris. There they were enthusiastic and together we fantasized about the many possibilities of the laser in solar cell fabrication. In March they will come to Amsterdam to sign a contract. Today, Jaap and I had to promise Philips Eindhoven in writing never to work on deposited layers again and instead to do a decent systematic study on laser-annealing of implanted silicon and the influence of various dopants and the wavelengths of the laser. What we do not value in the Netherlands are the mountains of excitement and fantasy. From fear of altitude and lack of bright ideas you have to proceed here through the polder of systematics with its straight canals and roads and a strong head wind as the only problem. Thinking of Holland/ I see conservative physicists/ slowly moving through/ endless systematic searches.

Now, reading through my diary on the early phase of our research, I start to understand what was happening. I wanted to be with them, with Bell, IBM, Stanford, Cornell and Philips. That is why this research was chosen. Having made this decision I surrounded myself with a facade of publicity, patent applications and research proposals.

It is a miracle that in this phase a basis was laid for a PhD thesis and a collaboration on solar cells. This is solely thanks to the sober farmer's instincts around me.

Let us see which effect our systematic research finally had.

New York, 13 October 1978. I am on sabbatical at the IBM Research Centre in Yorktown Heights with Leo Esaki, where I am having a great time. Without a family one has another eight hours extra. During the day I analyze Leo Esaki's superlattices and during the night I work with Jim van Vechten on laser-annealing. He is a typical example of an American workaholic, who knows everything of the thermodynamics of semiconductors. Yesterday, I read a paper by Rimini in Physical Review Letters. They see copper atoms after implantation and laser-annealing move all the way to the surface and this, according to Rimini, is proof of the thermal melting model, because they can calculate the copper distribution exactly using the equilibrium segregation coefficient of 10^{-4}. Dick Hoonhout, however, has measured the profile of at least ten different species with segregation coefficients varying from 0.3 to 10^{-8}, and he does not detect any correlation. So, the agreement for copper the Catania group finds must be fortuitous and not at all proof of the melting model. In the afternoon I say this to John Poate of Bell Labs who happened to be visiting IBM. As if I had stepped on his toes he jumped up and presented five different proofs for the melting model. Fortunately, Jim van Vechten came forward and dismissed John Poate's proofs one by one. The result is a very excited atmosphere, a heated debate of physicists yelling but not listening. After the smoke has lifted there are clearly two camps and a controversy is born.

Amsterdam, 4 May 1979. Jim van Vechten visited us this week on his way back from a conference in Tbilisi. He carried a draft of a paper with him and asked Dick Hoonhout and myself to be co-authors. With staggering amazement I read through all forty pages. It was a clear example of a non-scientific paper. Jim discussed at least fifteen different experiments of others, of which most were not even published. He kept a score card to show that none of the experiments agreed with the melting model, four contradicted heating and all fifteen could only be explained on the basis of ionization and plasma formation. Some time ago Jim had had trouble working at Bell Labs and he had moved to IBM, and the paper was full of citations, characterized as 'stupidities' from papers out of Bell Labs.

Tuesday morning I asked Dick, who had read the paper also: 'What should we do?' Jim comes in and Dick says very cool and straight: 'If a theory is in agreement with fourteen experiments, but in disagreement with number fifteen, then it is true the score is fourteen to one, but the theory has effectively been proven wrong. So, all we need to do is to choose from all the arguments the strongest one, to write this down as clearly as possible using only facts from the open literature and publish this as quickly as possible as a letter.' Thus we set out to do so and spend the rest of the week writing an article under the title: 'Reasons to believe pulsed-laser annealing does not involve simple thermal melting', in which we give five solid experimental facts that contradict the melting model. This result is celebrated with a trip along the river Vecht, the native soil of Jim's ancestors. On our way we are afraid that our paper will never pass the referees from Bell Labs and we decide to send it to Joe Budnick, a former IBMer who is also editor of Physics Letters. In addition, we will present our arguments at the Gordon Research Conference this summer and at the MRS meeting in Boston in November. In principle we also agreed to write a second paper on laser-ionization effects and plasma, but we had no time. Fortunately for me, because that part of Jim's story I did not understand, but you cannot say something like that out loud.

Analyzing these citations more closely the following appears to be happening. First a table of experimental results comes to life because a few Italians stick their neck out: 'their proof of' makes us claim 'no proof of'. Then the meaning of the same experimental results changes again and it says: 'in contradiction with'. The latter statement primarily results from personal resentment elevated to scientific controversy. They believe this, so we believe the contrary. This is also much more exciting than just going along with them.

What will come of the scientific method of falsifiable hypotheses and experimental verifications we may see in the following parts of this diary.

New York, 28 July 1979. It is five o'clock in the morning. Just returned to my hotel. Worked happily all night. At the Gordon Conference our critique of the melting model was well received, yet hardly anybody was willing to put it aside. Therefore, after arriving at IBM we have immediately started writing an article on the plasma model. Spent the whole night behind the computer together with Jim van Vechten and Ellen Yoffa, a post-doc who does calculations on laser-ionization and heating effects.

When the draft was finished I asked the computer how many different words from the dictionary we had used to write our article. The answer was: a vocabulary of only 342 words. So our shabby and petrified research efforts are reduced to dry stones of knowledge before being accepted by the scientific literature.

Amsterdam, 3 December 1979. Dick returned from the MRS meeting in Boston. That silicon melts under the laser pulse is apparently generally accepted. The disagreements we have noticed would easily be attributed to the non-equilibrium conditions during the fast solidification process. Is that so? We will not get our contribution to the proceedings of the Boston conference published if we do not add such a statement to our paper. Dick says: 'I don't know if you still want to be member of the counter-group, but I don't.'

The scenes just described do not seem to give any support to the classical picture of scientific progress via the interaction between falsifiable hypotheses and experimental verifications, or via the organized scepticism against new scientific information regardless of career and reputation of those concerned. On the contrary, we appear to have to do with an example of the theory on the role of paradigms. The laser-annealing community has as soon as possible embraced a paradigm, the melting model, in order to collect and explain all experimental results. Attacks on this model are made impossible from the very beginning by not putting up the theory for verification but as rules of the game. It is funny that behind the 'iron curtain' the rules of the game of laser-annealing are played with ionization and plasma, whereas in the West thermal melting is the password. Because communication is poor, this is hardly a problem. It illustrates what is also apparent from my diary: paradigms come about via psychological and social factors such as the desire to be a member, loyal, gregarious. If you want to become famous by creating a revolution, which means changing the rules, you will have a hard and lonely way ahead.

Next, in my diary on laser-annealing, follow a large number of empty pages. During this period Dick Hoonhout writes a computer code of the melting model. He knows in advance that the agreement with his experiments will be perfect, for there is a sufficient number of free fitting parameters to use. In the summer of 1980 he writes his PhD thesis in which he presents a systematic study on crystal growth of silicon with the laser, but on the physical mechanism, the fundamental problem, the thesis is not conclusive.

In September 1980 I attend a big international conference on the Physics of Semiconductors in Japan.

Kyoto, 10 September 1980. I am sound asleep. The telephone rings: 'Hi mate, you better rush, you are session chairman in ten minutes.' It is Ian Mitchell who saves my bacon, the man whom we laugh about because he is always late.

Is Jim van Vechten right after all? Today he has introduced me to Compaan while saying: 'Here is yet another Dutchman who is in for a fight.' Compaan produced time and wavelength dependent reflectivity measurements of the silicon surface indicating that it certainly does not melt during the laser pulse. His results have already been accepted by Physical Review Letters but not yet by the laser-annealing community.

When I asked Walter Brown why nobody of his group would try to reproduce the results of Compaan, Walter answered: 'There is nothing to gain. If Compaan is right, he will get all the credit and if he has made a mistake everybody will say "we knew it all along"'.

What should we do with that? I do not know enough of optical measurements to judge the reliability of his work. In the laser annealing session Jim van Vechten gave a terribly complicated theoretical talk and afterwards asked me what I thought of it. I only said that I still thought it an interesting controversy.

The final section of my Diary of a Physicist I wrote in Bombay.

Bombay, 13 February 1981. Every day, in the morning and in the evening, I see the most horrible, appalling scenes for a full hour as we drive in our little bus from the Tata Institute for Fundamental Research where I stay, to the Bhaba Atomic Research Centre on the other side of this city of beggars. There a conference is held on ion implantation and laser-annealing.

In the bus the subject of discussion is of course laser-annealing. Our bus has to stop for a red traffic light. A man shuffles towards me, he has only one arm, between the rows of cars he comes to me. In front of my window he lifts his one arm. Instead of his hand I see a sharp little stump. If I wanted to give him any money he would not be able to accept it, I think. Is that the reason I shake no to him? The bus starts moving again and our conversation, on the controversy in laser-annealing, results in a new idea for an experiment. Was that the reason for me to stay on the bus?

A diary is a very personal document, in which you preferably write down what may not be discussed in public. Yet I have chosen to read from my diary instead of presenting a speech, because the ceremony of a speech does fit with my intentions to get rid of a myth. The myth in which the question of legitimacy of science is answered like in church: 'Why science? To satisfy our curiosity and therefore be happy here and ever after.'

Knowledge is not discovered by researchers on their path through nature. Scientific knowledge is created most literally. During this creation not only does curiosity play a role, but also prejudice; not only rationality, but also belief and emotions; fashion; pride and glory; friendship and jealousy; fanaticism and mental laziness. Of course these are not specific properties of physicists in the ion-beam community. In the diary of every physicist you may read stories such as I have cited here.

As I have said earlier, I believe it is harmful to ignore the human factor in our activities as scientists. The harmful part lies in the specialization in only one direction, without interaction with other dimensions. The great academic problem is that we have specialized so much away from everything and everybody that our work is no longer related to anything. So that we fall silent when we come home in the evening and are asked: 'Darling, why are you so late?'

It is harmful for young physicists to get into a process during their education in which norms and ideals are changed, because they learn to forget and suppress the emotional basis of those norms and ideals. This atmosphere, which is conserved because of a myth and perhaps may best be compared to the training of priests in the Roman Catholic Church not too long ago, has led to generations of scientists who do not want to be bothered by hunger and poverty, the energy crisis, economic depression, materials shortage and environmental disasters; who do not want to be held responsible for the influence of nuclear power and weapons, computers and telecommunication in our society.

My view on the social responsibility of the scientist is that he should be fully aware of all signals, factors and forces working on him and he should make a choice conscientiously. Which choice? That is up to the scientist. But before socially responsible science is possible, we should be free enough at red traffic lights to get off our own little bus occasionally.

1981

Dear Zhong-lie

Suppose a nuclear physicist who is working with low energy accelerators wants to do something else with his expertise. What do you think is the probability that he will find a new subject and be productive in a few years time? Let us estimate this at 10 per cent.

Suppose two scientists meet at a conference in Denmark. One is from the People's Republic of China, the other from the Netherlands. They share a common interest in pulsed laser annealing of silicon and would like to collaborate. What do you think is the probability to find this Chinese scientist working in the Netherlands the following year: 10 per cent?

Suppose a scientist, who has just arrived from Beijing, wants to go from his apartment in Amsterdam-Noord to the Uithof in Utrecht by public transportation. He has never travelled outside China. Do you think he has a 10 per cent chance of arriving at the Van der Graaff Laboratory before coffee time?

Suppose a physicist starts working in the field of particle solid interactions. What do you think is the probability that two years later he will have published seven scientific papers in internationally refereed journals: 10 per cent?

Suppose a Chinese physicist wants to get his PhD in the Netherlands, but he has no Dutch graduate degree so he has to ask the Minister of Science and Education for permission. Will he have a 10 per cent chance of getting a positive answer within a year?

Suppose a man who cannot cook starts living in a foreign country and has to look after himself for the first time. What is the probability that he will not become ill in two years' time: 10 per cent?

Suppose a physicist working on laser annealing of silicon, after one year switches to ion beam mixing. What is the probability that his thesis will be finished in two years' time and will contain an equal number of chapters on both subjects: 10 per cent?

A Chinese physicist arrives at a Dutch laboratory with a complex computer programme he has written in Beijing. What do you think is the probability that the computer in Amsterdam will produce reasonable data within a few months: 10 per cent?

A group of physicists starts working in a new field. Do you think that within two years 10 per cent will have presented results of their research work at two international conferences and given seminars in six different laboratories?

What fraction of university professors who do not have a PhD will go through the trouble of still getting a doctoral degree, 10 per cent?

Now we can answer the last question: On September 1980, a Chinese citizen arrives at Schiphol airport in his typical dark blue high-collared suit. What is the probability that on 13 September 1982, he will wear a black and white suit to receive his doctoral degree?

Multiplying the above probabilities one arrives at 10exp-9 or one in a billion!

Fortunately, the total population of China is one billion, so thanks to your unique perseverance and determination, your ability to adapt yourself and your qualities as a theoretical and experimental physicist, the event has occurred!

I am sure to speak also in the name of professor Kistemaker and all my colleagues in Utrecht as well as in Amsterdam when I say that we find it an honour to have been involved in this historical event.

My congratulations and I am looking forward to the continuation of our collaboration.

1982

Alchemy

Everything we see in nature consists of a combination of matter from the periodic table of elements. In the nineteenth century the Russian scientist Mendeleyev discovered how they should be ordered, from light hydrogen, number 1, up to heavy lead, number 82. In the meantime nuclear physicists – actually they are nuclear chemists – have added quite a few elements. Up to uranium, element number 92, they may be found in nature. Super-heavy elements are not stable and have to be created in the lab. They don't have to be of course, but if something is not there which could be there, an interesting question is why it is not there. At least, physicists think so.

In various places in the world accelerators are used to collide heavy nuclei with each other, in the hope of fusing new elements. In this way modern alchemists have successfully filled the open spaces in the system of Mendeleyev one by one . Especially the professors Seaborg and Giorco from Berkeley were leaders and created the elements 95 Americium, 97 Berkelium, 98 Californium. The Russians could not stay behind and professor Flerov in Dubna created element 101, which he baptized Mendelevium. The Americans earned the Nobel Prize for their work and celebrated this with the following product, element 102, Nobelium.

Since then German physicists have joined the race with the super-heavy ion accelerator in Darmstadt. It was especially built, because it was predicted that element 114 and also 124, once produced, might be stable enough to remain. As the atomic number increases, so does its radioactivity. The super-heavy elements emit alpha particles and thus lose weight. For element 106 that process went so fast that when professor Flerov thought he had made it, the Americans did not believe him because he could not collect enough material to prove it. But Flerov tried to cheat: he went with his results to the Germans and suggested to call his new element Hahnium, after the German nuclear scientist Otto Hahn. This proposal was not immediately accepted by the International Union for Pure and Applied Chemistry.

In the meantime, the Germans had also laid a claim there on element 109, of which they had produced no more than a single atom. Their results were received with some considerable disappointment. Not only because of jealousy, but also because from the experiments in Darmstadt one had to conclude that it is virtually impossible to extend the periodic table of elements further. The result of the violent collision between two very heavy atomic nuclei is merely that they fall apart into many fragments rather than fusing together. For the moment element 109 will be the end of Mendeleyev's table and the nuclear chemists still remain with a big controversy over the names of the elements 105 through 109.

Yet, atomic physicists already know a lot about super-heavy elements with an atomic number larger than 109, even though they may never come into existence. Many properties of the atom are determined not by the nucleus but by the electron cloud around the nucleus. The atomic nucleus determines the mass of the atom and its positive charge by which the electrons are bound. The electrons determine the chemical properties of the atom and are also responsible for magnetic and optical properties of matter.

Atomic physicists differ from nuclear physicists in that they only study the electrons, not the nuclei. These electron-physicists also have collided heavy atoms onto each other in order to simulate super-heavy atoms. During a collision of two super-heavy atoms both nuclei may come so close that to the electrons it seems as if there is only one single nucleus, with a total charge equal to the sum of the two colliding nuclei. If for instance the atom of element number 32, Germanium, collides with atom number 31, Gallium, then the electrons don't know any better but to form the electron cloud of the atom 32+31=63, Europium, during the collision. Such a united atom can indeed be observed, as Dutch atomic physicists have discovered by detecting the atomic radiation that may only be observed during the collision. If you collide two uranium atoms with each other you should observe radiation from atom 92+92=184! Consequently, atomic physicists have in the meantime observed the characteristic radiation of all elements between 109 and 184, whereas these elements have never actually existed and according to some nuclear chemists never will exist.

These questions remain extremely interesting for physicists.

1984

Worthwhile

Sometimes it happens that I am asked at the end of a laboratory visit, especially in Third World countries, what kind of research should be done with the facilities that were shown. My answer always is: 'You shouldn't ask me what is or isn't worthwhile for you; you know that best.' After some insisting I may add: 'You don't want to do exactly the same research we are doing, do you?' Recently, when that also didn't work and my hosts began to press me, I tried to tell them what my criteria are for my own scientific choices. I happen to have three criteria and the nice thing is you can measure to what extent my chosen research fulfils those criteria, test it explicitly, yes even quantitatively.

The first thing I always ask myself before starting a research subject is: What is new? Every idea, big or small, may be summarized in a few words, in which the answer is given to the question: Suppose the work is successful, what did we learn from it and what did we accomplish? The answer can be staggeringly simple, but if you can't tell you don't know what you are doing. The minute I hear myself say that something is too complex to explain, I smell trouble.

The news value of the expected results can be measured by answering the following question: Where will the results of this investigation land eventually? In case the answer is in the *Physical Review Letters* (the most prestigious journal in physics) then it is OK. Unfortunately, I don't have such ideas very often. No worries, the answer may also be: in another international science journal or at an international conference. But if that is not the answer and a patent will also be excluded, it is probably wiser to put the idea out of your head. There is always something wrong with whatever idea it is (otherwise we didn't need to do research). Frequently one finds something different from what was expected, but that is precisely what makes research so exciting. An idea may be ever so good, but if you know beforehand that nobody will be interested, you are casting pearls before swine.

My second criterion is a matter of time. How long will it take before the specific research will produce results: a week, a month or a year? Ideas that within a week lead to publication in Physical Review Letters are the best ones, but they are scarce. Even when in practice it turns out that it takes three weeks, still nothing wrong. But if you think that a research subject will take a year's time before anything may be published you should take into account that this estimate also may be too short by a factor of three. In that case it is perhaps better to admit that you are not (yet) ready for the specific idea, and that a less ambitious plan should be preferred. Moreover, good ideas usually occur in several places at the same time 'when it is that time of the year'. Whoever needs more than a year in such a case will absolutely be too late. Other research groups will have the scoop. An idea may be brilliant, but if elsewhere one is better equipped for that specific research it is better to have it done elsewhere. By the way, this is an excellent reason for international co-operation.

Scientific research costs a lot, first and foremost for the scientist doing the research. In addition, you can't do the research all by yourself, certainly not in physics, so you ask from the people around you not only understanding and patience, but also assistance and from funding agencies a rather large quantity of money. Reason enough to doubt whether a research proposal is really worthwhile. This is my third and most important criterion. How do you know if something is or isn't worthwhile to investigate? How can you possibly measure that? By trying to find the people to do it AND the money.

1984

Scientific nomads

Floating on the Long Island Expressway in my rental car, there isn't much that gives me such a kick. The feeling of freedom after having been locked up in rows of people: in front of the KLM check-in, waiting for customs, inside the airplane and customs again. The radio blares the latest hits while I drive through the New York hills on my way to a former co-worker who has left to find his way to Brookhaven National Labs to go and work there. He lives in a country mansion in a park close to the sea, not even an hour's drive from Kennedy Airport. In the landscape where Woody Allen films in case the scene is not set in Manhattan, this Dutch physicist rents a huge summer mansion from a millionaire from Brooklyn. Close to the kitchen door is a baby monitor so that his wife can hear their eight-month-old daughter. 'This is quite different from our little apartment in Duivendrecht,' she says, while he is out in the snow chopping wood for the open fireplace, 'but you must have two cars.' I have come to ask him to become project leader in our laboratory in Amsterdam, but I am afraid he and especially she will decline the offer.

During the 1950s and 1960s Brookhaven National Labs was the centre of the world of nuclear and high-energy physics. In the meantime, these physicists have moved elsewhere and the National Synchrotron Light Source is situated here. My host takes me to the most intense light source in the world. Actually there are two sources of light: one for x-rays and the other for ultraviolet light. Originally these electron accelerators were built for nuclear and high-energy physics; today they are especially constructed for atomic and molecular physicists, chemists, biologists, medical researchers, materials scientists and engineers. In Brookhaven the source of light is surrounded by at least sixty different experiments of some hundred researchers spending shorter or longer periods here.

I visit experimental facilities of all large industrial and national laboratories all over the world. Exactly the same kind of equipment might have found a place in Amsterdam. A few years ago a proposal to that effect was made, but this was way past what Dutch imagination could handle. Our

Philips may be the world's biggest producer of light, but for invisible light we depend on facilities abroad.

Before the sun sets into the ocean in a flaming red sky, we walk along the beach. With great enthusiasm my host talks about six experiments he has already been able to do in the past year. He also tells about his plans for next year. Even the wind and the waves seem more spectacular here than in the North Sea; their roaring makes further conversation impossible. Back home I ask him if he would like to become project leader of our synchrotron experiments, first in England and then later also in Grenoble. 'This sounds like music to my ears,' he says, 'in a few years Brookhaven will be deserted.' The 'nomads' will be leaving for the next source of light, which is already under construction in the neighbourhood of Chicago. By that time he would like to go back to Europe. His wife says: 'I would like to stay here but I will go with him.'

In our newspaper Henk Hofland wrote: 'The entrance to the Picasso and Braque exhibition costs $7 plus the price of an airline ticket.' Before going on to Boston I go to the Museum of Modern Art; travelling scientists get a reduction on the airline ticket. Hofland is right: this exhibition is worth the money even without reduction. In vain I try to predict which painting is Picasso's and which Braque's. Yet in those days Picasso scooped up the prizes, Cubism was associated with him and Braque felt abused. Before leaving the museum I discover Charlotte van der Waals' folding vases. Apart from Rietveld, here we have another Dutch designer among the world's greatest. Folded flat these vases can be delivered through the letterbox. Unfolded they belong to the most beautiful flower vases I have ever seen. Charlotte very elegantly solved the problem if you want to bring not only flowers but also the vase to hold them.

The annual meeting of the Materials Research Society in Boston begins and all nomads are present again. Two and a half thousand participants, divided over just as many symposia as there are letters in the alphabet. The programme book of this meeting is thicker and heavier than the telephone directory of Amsterdam: three thousand scientific contributions, an exhibition of instruments and books. It is all about new kinds of steel, electronic application of diamonds, liquid crystals, polymers, high-temperature superconductors, new ceramic materials, changing of properties of materials by means of laser beams, particle accelerators and electron beams; all together in two huge hotels. The conference starts every morning at eight with papers being presented until late in the evening. For lunch one can eat a bun and watch TV programmes in which modern developments in materials are being presented for the general public and distributed over the US by satellite. All very different from the average scientific meeting in the Dutch retreat houses, where one can't start before coffee time and where the participants definitely have to catch the train at tea time otherwise they will be late for dinner at home.

We are here with seven PhD students and two group leaders from our lab. Like rich gypsies we stay in two hotel suites for $30 pp/day. We will present 12 conference contributions between us.

In order to hear the situation of high temperature superconductors I listen to the Nobel Prize winners. The room is packed; certainly half of the people present have only just started working on superconductors. Most of them come from entirely different fields, such as nuclear or atomic physics. Enticed by the glamour of new discoveries, but also by the green pastures of a new field of research, in which one hopes to contribute one's own expertise in making and analyzing high-Tc superconductors of atomic scale. The best superconducting layers are made by means of a laser beam. A method discovered by a Dutch atomic physicist, now working for Philips but then still at Bell Labs New Jersey.

At another symposium, an American Dutchman surprised the competition by using arsenic as soap on a silicon wafer and in this way getting beautifully smooth layers of germanium on silicon. Thus one hopes to make a new laser beam for optical communication by means of glass fibres. During the coffee break I congratulate him and inform him about some vacancies at Dutch universities but he hardly seems interested. For the past six years he has been working at the IBM lab in New York where he has his own group now. An Australian Canadian joins us. He has been living in Canada for twenty years and has just recently rejected a most attractive offer from Australia. 'No,' he says, 'I can't first let my parents down and then also my children.'

In the afternoon, four of us present our work at Varian/Extrion. Five years ago on the dot I got to know this business because the Dutch director of ASM Int. was brave enough to invest in it. In three years' time a new machine for the computer chip was created. Last year this piece of Dutch technology was taken over by Varian for a lot of money. We still have a co-operation contract with them and report regularly. I sincerely hope the contract will be continued. They are interested in our results, as they should be, because my boys have been working very hard. During the discussion it becomes clear that they indeed want to continue, for new research suggestions are being made.

At four in the morning the fire alarm impels the hotel guests to the corridors in their boxer shorts and dressing gowns. False alarm, but I cannot sleep any more. In a few hours' time I will have to do a presentation. Even though I like doing that I am still nervous. I have learned by now that it is a good sign: without stage fright no good performance. We think we can predict the temperature needed to grow amorphous materials efficiently without crystallization. My story sells like hotcakes. I have seen different times, especially at this conference. Have they finally given in? I cannot believe it.

We have dinner with old friends who all worked at the FOM Institute in Amsterdam at some time or another. In addition to my own group there are two men from India who are now living in America, three Dutchmen, of whom two live in the US, one Chinese and five Japanese. The latter always go back to their homeland. One of them asks me: 'Why on earth have you gone back to the Netherlands after living in Canada and the US?' I tell him: 'Because my roots are there, I have never been able to cut my roots.' I think there is more: my present job is the one I have always aspired to and 'in the land of the blind the one-eyed man is king', but in America there are so many excellent people; moreover, the scale of the laboratories is much larger and I don't like that. The Dutch Organization for Scientific Research is a paradise compared to the National Science Foundation. And I would never hold up my hand to the military to get money for research. Why should I teach American students instead of Dutch? There is more: I feel handicapped in a country where I cannot express my emotions in my mother tongue. It is so easy for children, but parents never learn that anymore.

It is time for the other group members to present their work. In three different sessions they all stand in the spotlight in a hall or with a poster board, which nowadays furnish the proverbial lobbies of international meetings. It is quite something to see these graduate students, as they are called, at work. Two years ago we put a new accelerator into use in our lab; the results of the entire research work must be put together in a message of fifteen minutes at most. Without exception they manage to do this, which results in discussions with their peers from all over the world, giving them new ideas for new experiments. And a young American presents himself; he would like to work as a postdoc in Amsterdam.

At the end of the conference we pack our suitcases. Four of us go to Bell Labs where one of us has started as postdoc recently, one goes to Harvard to finish an experiment and others go and visit friends or relatives in New York and California. I say goodbye to my colleagues here. Some of them I have known for years and they have become my best friends. Even though they live on the other side of the globe we, scientific 'nomads', meet each other several times a year. If it didn't involve so much travelling I would probably never have become a physicist.

1989

Superheated Ice

Nature is not fair to people who like skating. They are told to stay away from first-night ice. Worse, they have to wait days until the ditches are cold enough for the ice to be safe, yet as soon as thaw sets in the skating is finished. Nature is strangely asymmetric: just like ice, most materials immediately start to melt as soon as you heat them to their melting point, but as you cool them down again most materials stay fluid until well below their freezing point. Materials can be under-cooled, but they cannot be superheated. Superheated ice is unheard of, except in the crowds at the World Championship Speed Skating. In nature superheated ice does not exist.

Melting is a common phenomenon. Yet the mechanism of melting on the scale of atoms or molecules is still not properly understood. During the melting of a material there are two phases present at the same time. When we heat the material until it melts somewhere in the solid phase the liquid phase will set in. If we continue adding heat the temperature will not rise anymore, but the fluid phase takes over at the cost of the solid phase and finally there will only be liquid. According to Lindemann, materials melt as a result of the heat-motion of the atoms in those materials. In 1910 he already discovered a connection between the heat-motion and the melting point of a material. With high temperatures the atoms more or less shatter the solid into pieces and thus, according to Lindemann, the crystal structure is lost.

If this were true it is hard to understand why materials are not shattered to pieces completely. Why should the solid and liquid phase co-exist at the melting point? At the melting temperature it is as if the fluid has to nucleate somewhere and then grow further. This indicates, according to some physicists, that at the nucleation sites an extraordinary concentration of crystal defects is created. One may think of vacancies or interstitial atoms. However, the experimentally determined concentration of such defects is so low that they cannot possibly play a serious role. It is more probable that melting sets in by a swelling of dislocations in the crystal. As a matter of fact the liquid phase is seen as a highly disturbed solid phase.

The greatest disturbance of a crystalline material is where the crystal ends, at the surface. Because the atoms at the surface are missing their neighbours above, they are less tightly bound than the atoms inside. That is why they readily move from their normal positions. Melting may then happen as follows. The weakly bound atoms at the surface are the first to leave their positions upon heating of the solid material. In this way, already below the melting temperature a disordered layer is formed of a few atoms thick. As the melting point is approached the thickness of the disordered layer increases until the crystal at the melting temperature melts in its entirety from the surface into the bulk.

Faraday found the first indications for surface melting in 1842, in a series of experiments with blocks of ice. Considering the ease with which two ice-blocks touching each other grew together, Faraday deduced that the ice had to be covered with a thin layer of water. When the blocks make contact the layer is surrounded by ice on both sides. The atoms on the interface are locked, lose their mobility and freeze. Thus the two blocks are cold-welded.

Since Faraday's experiments a lot of research with a variety of materials has suggested the existence of the phenomenon of surface melting. However, there was no concrete proof in any of these experiments in the sense that the melting layer itself was never detected. The fact is that the layer is only a few atoms thick and requires atomic physics detection methods to be made visible. In 1985 J. Frenken and J.F. van der Veen of the FOM Institute in Amsterdam bombarded the surface of a lead crystal with a beam of fast protons. In the signal of the reflected protons pronounced shadowing effects occurred. The atoms in the surface layer of the lead crystal were perfectly in line and shadowed each other. As soon as the atom layer began to melt the shadow disappeared from the surface, whereas in the signal from the bulk of the crystal the shadow remained clearly visible. In this way it was possible to detect the melting thickness of a single atom layer.

In the meantime a kind of international 'melting school' has been established of physicists and chemists who study the melting of surfaces under all sorts of conditions. It has been found that the mobility of atoms in the very top layer is much higher than in the bulk of the crystal. Another discovery is that surface melting does not occur for all crystal surface orientations. The thickness of the molten layer does not only depend on the temperature, but also on the crystal orientation of the surface.

Melting as well as freezing both need a seed. The ice-masters of the famous Eleven City Race in Friesland have the habit of floating pieces of ice into the blowholes under the bridges to speed up the forming of safe ice that can carry the speedskaters. When melting a material there is usually a seed present on the outside, the very surface. Since we know how melting starts, we also know how to prevent it from happening. Unfortunately, this will not be of any use for people who love skating. By covering the surface atoms with a different material that has a higher melting point, the atoms at the interface are kept in place. In this way it is even possible to heat solids high above the melting temperature and postpone melting. This shows that melting is not just a matter of shattering a crystal to pieces. The thermal motion of the atoms in a superheated crystal will be larger than their motion near the melting point. Yet the crystal will not melt as long as there is not a seed at very the surface.

We know where melting begins, but we still do not know why. It is unknown to which temperature solid matter can be heated and also whether there really is a maximum temperature. It is true what Professor Van der Veen said in his inaugural lecture at Leiden University: 'One is likely to think that physicists, who are most familiar with the natural phenomena, should know by now how a process like melting takes place. I have to disappoint you. The physicists have collected an impressive quantity of knowledge on the building blocks of matter, but about how those blocks join together the physicists are dumbfounded. Let alone that they can explain why the cohesion fails upon heating and the blocks start to move about. The physicist of this day and age can tell interestingly about quarks and the first three minutes of the creation of the universe, but he cannot answer the question why butter melts when we put the pan on the fire.'

1990

Managing a discovery

There is widespread belief that scientific research cannot be planned and that managing a discovery is in contradiction with true exploration and the real nature of the explorer. This belief, however, is based on a misunderstanding, as will be pointed out through a historical example of a major exploration to illustrate the management qualities needed for a discovery and how they are best developed.

On 16 August 1880, Robert E. Peary wrote to his mother: 'I do not wish to live and die without accomplishing anything or without being known beyond a narrow circle of friends. I wish to acquire a name which shall be an "open sesame" to circles of culture and refinement anywhere, a name which shall make my Mother proud and which shall make me feel that I am peer to anyone I may meet.' Twenty-four-year old Peary had set his mind to voyages of discovery in the area of the North Pole. However, he was offered the leadership of an expedition of at least a hundred men straight through Nicaragua; the aim was to explore the route for a channel connecting the Atlantic Ocean with the Pacific. Peary accepted this honourable mission, but before completing this job he heard the news that the Norwegian Nansen had crossed the ice cap of Greenland from the east to the west coast. Immediately Peary stopped his work in Nicaragua. Although he had entirely worked out the plan for the channel, he did not wait for a decision on its implementation. Apparently he had a good nose for what was worthwhile and probably also an inkling that the new channel was going to be in Panama.

Peary made a new plan for a voyage of discovery in North Greenland, where the map still showed large white areas in those days. He was granted leave of absence as civil engineer for the American Marines with full pay, but on the condition that he would pay his own expedition and would not claim compensation in case something should happen to him. This was the beginning of a period of twenty-three years making expeditions and mapping out vast areas of Greenland and Northern Canada on the way to his final goal: the North Pole. His trips always started by boat with a small crew, carefully selected by himself. Assistants who could be counted on to be able to lead the many Eskimos who were taken on board during the trip together with their wives, children and dogs. At first Peary had thought he could reach the North Pole using sleds on the ice cap covering Greenland as an 'arctic highway'. But Greenland turned out to be an island, and also the northern part of Canada did not extend further north than 400 miles from the Pole. Here he sailed in the summer; the autumn was used for hunting and fishing in order to build up enough provisions for the long winter. In early spring, when it became light, the real voyage of discovery began with a few white assistants, tens of Eskimo's and an equal number of sleds pulled by more than a hundred dogs.

The expeditions cost $60,000 to $80,000 each, in those days a lot of money, which Peary managed to collect in all sorts of ways. To begin with he gave hundreds of lectures all over America dressed as an Eskimo appearing onstage with a sled pulled by a team of loud barking dogs. In this way a series of 169 lectures brought in $20,000. Prominent American millionaires became members of the Peary Arctic Club for $5,000 per person. One of his most important supporters was President Roosevelt, who compared Peary with Columbus and Magellan, and after whom Peary named his ship. During one of his journeys he discovered an enormous meteorite, hoisted it on board and sold it to the Natural History Museum in New York, which exhibited the metal lump together with six Eskimos whom Peary had brought in passing but who died of pneumonia during the exhibition.

The expeditions had to find their way over the pack ice of the Northern Ice Sea. The Eskimo sleds were less suited to this and in due time Peary developed his own sledge, with handles on the ends and longer gliders to make it possible to carry them across the drifting ice and crevices. Pack ice moves continually due to the current and wind at sea. In his reports Peary describes how he tries to sleep in his igloo after a day trip while the ice is breaking all around him. He does not lie in a sleeping bag because he is afraid that the ice under him might break and he will drown. Sometimes they have to wait for large cracks of open water until they have frozen again and the ice is thick enough to support the entire group. It is life-threatening if the ice breaks behind the front group and they are cut off from supplies. This forces Peary to stop his expeditions several times, in order to prevent death from starvation. Thus he has to make a detour back to his ship. Usually there is not enough time left for a second trip to the Pole before the winter. In order to prevent a complete failure the remaining time is used to map the coastline further and to name interesting locations after the sponsors of the Peary Arctic Club. On every trip Peary manages to penetrate the Arctic further north, until on 6 April 1909 he is the first to reach the North Pole.

Today the North Pole is no longer an area to discover, in spite of the fact that many people continue to make 'expeditions' there. On our globe hardly any white spots are left, but in the scientific world there still is plenty to discover. Let us view science as a voyage of discovery. Managing a discovery demands on financial and economic skills from the discoverer, also personnel management is indispensable because these voyages are teamwork, also in the realm of science. Moreover, it is absolutely necessary for the leader of the team to be a fanatic discoverer. In the 1960s the universities were democratized and scientific research came into the hands of councils, a diffuse group of people who soon lost track. In the 1980s managers were appointed, people who saw no difference in running a university, an electronic industry or a hospital. People who think that management is a profession in itself, which you can learn at a university or better still at business school. Nobody will ever think of asking a student from a business school to lead a Himalayan expedition or a journey to the North Pole. Why then the management of a research organization or a university?

How do you manage a discovery? When is an expedition successful? Take Peary, who in his book 'The North Pole' gives a list of 'Essentials that brought success':

'To manoeuvre a ship through the ice to the farthest possible northern land base from which she can be steered back again the following year.

To do enough hunting during the autumn and winter to keep the party healthily supplied with fresh meat.

To have dogs enough to allow for the loss of sixty per cent of them by death or otherwise.

To have the confidence of a large number of Eskimos, earned by square dealing and generous gifts in the past, so that they will follow the leader to any point he may specify.

To have an intelligent and willing body of civilized assistants to lead the various divisions of Eskimos – men whose authority the Eskimos will accept when delegated by the leader.

To transport to the point where the expedition leaves the land for the sledge journey beforehand, sufficient food, fuel, clothing, stoves (oil or alcohol) and other mechanical equipment to get the main party to the Pole and back and the various divisions to their farthest north and back.

To have an ample supply of the best kind of sledges.

To have a sufficient number of divisions, or relay parties, each under the leadership of a competent assistant, to send back at appropriate and carefully calculated stages along the upward journey.

To have every item of equipment of the quality best suited to the purpose, thoroughly tested, and of the lightest possible weight.

To know, by long experience, the best way to cross wide leads of open water.

To return by the same route followed on the upward march, using the beaten trail and the already constructed igloos to save the time and strength that would have been spent in constructing new igloos and in trail breaking.

To know exactly to what extent each man and dog may be worked without injury.

To know the physical and mental capabilities of every assistant and Eskimo.

Last, but not least, to have the absolute confidence of every member of the party, white, black, or brown, so that every order of the leader will be implicitly obeyed.'

In spite of this long list of essentials Peary was a lot less successful than he had anticipated. When he returned from the North Pole he was asked to prove that he had actually been there. But how can you prove, in a big white world where nobody has been before, that you really have reached your goal? Dr Cook, a former assistant of Peary, even claimed to have reached the Pole a year earlier. It became known, however, that a picture Cook claimed to have taken of himself on Mount McKinley was forged. Yet this incident only made it more difficult for Peary to prove he was right. He asserted to have travelled straight north along the 70th meridian. But there were doubts whether he would have been able to keep his track in the strong wind and with drifting ice. Peary had a simple manner of navigating. At twelve noon he measured with his sextant when the sun was at its summit and thus stood exactly south. Then Peary turned around and looked at his own shadow, which at that moment was pointing exactly towards the North Pole. He did not bother to make sure how much he deviated from the 70th meridian to the east or west. Peary thought that was an unnecessary loss of time, all he wanted to know was which direction was north. Unfortunately, that is why he could not point out on a map which zigzag route he had taken on the polar ice. The explorer Amundsen did believe him and said: 'I know Peary reached the Pole, for I know Peary.' And Amundsen adopted Peary's way of navigating on his race to the South Pole, which made him arrive just a little earlier than Scott.

To this very day Peary's claim is still questioned. The Englishman Wally Herbert has calculated the probable course of Peary based on known movements of the ice in the Polar Sea. Herbert comes to the conclusion that Peary must have ended up at least thirty miles to the west of the North Pole. There-fore the American Navigation Foundation has reacted with a study, which shows that Peary could not possibly have followed Herbert's route. What saves Peary's claim is that he measured the depth of the bottom of the sea through blowholes. The measured depths do not correspond with the known profile of the sea bottom under Herbert's route, but are consistent with the depths under the 70th meridian. President Roosevelt, who hoped to discover land under the ice field, commissioned Peary's depth measure-ments. Even though no new land was found with these seemingly unimpor-tant measurements, they now serve to support the claim that 'the stars and stripes' were first to wave on the North Pole.

In his legacy the Navigation Foundation has also found a number of pho-tographs, which according to Peary were taken in the igloo camp on the Pole. From Peary's notes and his photos one may conclude that they must have been taken at very different times of the day. The lengths of the shadows on the photographs tell us the height of the sun. It turns out that for a full twenty-four hours the sun is at 6 degrees and 30 minutes above the horizon. This is possible only on the Pole where in the summer the sun does not set and where according to the Almanac it must have been the height of the sun on 6 and 7 April 1909.

Returning to the management of a discovery, for that we need:
- To have a fine nose for choosing the right scientific area
- To have the convincing power to get the plans accepted
- To translate your own ambitions into general interest
- To be able to select people and create a team
- To navigate diligently
- To work pragmatically and not to put on blinkers
- To smell opportunities for unexpected discoveries on the way
- To be not afraid of the cold even though your toes get frozen
- To learn from the mistakes you make
- To develop techniques which are specific for the field
- To find the balance between your own interest and that of others
- To solve unexpected bad luck
- To have stamina and be lucky
- To want to win

All this for only one goal. Not the honour, because that is always disputed. The only thing that makes it all worthwhile is: to really establish something, to discover something really new.

How does one become a successful discoverer? You cannot learn it at business school. How does 'management development' work, not for the lighting industry or for a hospital, but for discoveries? For Peary it meant falling and getting up again for a quarter of a century, making big mistakes over and over again, but also learning again and again and in that way progressing one step at a time. In his book, *The North Pole*, he compared it with learning to win a game of chess and with that I would like to end:

> 'It may not be inapt to liken the attainment of the North Pole to the winning of a game of chess, in which all the various moves leading to a favorable conclusion had been planned in advance, long before the actual game began. It was an old game for me – a game that I had been playing for twenty-three years, with varying fortunes. Always, it is true, I had been beaten, but with every defeat came fresh knowledge of the game, its intricacies, its difficulties, its subtleties, and with every fresh attempt success came a trifle nearer; what had before appeared either impossible, or, at the very best, extremely dubious, began to take on an aspect of possibility, and, at last, even of probability. Every defeat was analyzed as to its causes in all their bearings, until it became possible to believe that those causes could in future be guarded against and that, with a fair amount of good fortune, the losing game of nearly a quarter of a century could be turned into one final, complete success.'

1990

Spy in the lab

To my surprise I had to introduce Ivan Matveyich to his colleague Taranchev. Of course the two atomic physicists knew each other's work, but because Ivan worked for the Academy and Taranchev for the University the two had never met before, so I was told. A car with chauffeur was waiting for us outside the railway station and soon we arrived in a grubby coffeehouse behind the Kremlin. One of the few that open at 7 am and even here coffee would not be served before nine. Over cold tea in filthy cups and dry bread we discussed plans for the day. In the morning Taranchev would show me around in his laboratory, after lunch I was to present my colloquium, then another lab visit and in the evening a party at Taranchev's home. Ivan told me he had to go to the Academy and would meet me in our hotel at night. 'Ivan Matveyich,' I said, 'do come to Taranchev's house, so I can introduce you to the entire Moscow group.' Ivan grumbled but I insisted, not knowing the serious troubles this would cause my hosts.

The day passed according to plan and after dinner the bell rang. Ivan came in but even before I could introduce him to the colleagues present, Taranchev called out loud: 'Here is the spy, the spy from Leningrad.' Ivan Matveyich did not say a word. Embarrassed I tried to explain to Taranchev that I had known Ivan for years, that he had worked in our lab in Amsterdam, that he had been a fantastic host last week and now accompanied me on a trip to several laboratories in the Soviet Union. 'And yet he is a spy!' Taranchev screamed again. Leaving the party seemed to be the best thing to do. When I turned around Ivan had a huge bottle of vodka at his mouth and emptied it in one gulp. Minutes later he could not stand on his legs anymore. Taranchev put him to bed and said that in a few hours the spy should be fit enough to come with me to the hotel. So, I had to stay there but I did not understand what was happening to me. When the party came to an end Ivan was put under a shower, fitted into his suit and taken to our hotel half-dazed. Finally, in the car Taranchev explained.

Everyone who had a foreigner visiting him was obliged to write a report afterwards and for the Russian bureaucracy 'everyone' literally meant every individual. Every person who had met me had to write a report about my visit. All those reports landed somewhere on a desk and consequently they had better be identical, otherwise there might be trouble. That was why Taranchev had written a report about this day and all the members of his group copied it from him. He could not ask Ivan Matveyich to do this, so Ivan had to write his own report and that made him the spy. Ivan apparently had foreseen this would happen, so he got drunk deliberately. For Russian bureaucracy this meant he did not have to write his own report and that is how Taranchev had the opportunity to stick a copy of his own report into Ivan's pocket.

This traumatic event took place during one of my first visits to the Soviet Union about fifteen years ago. This summer I attended a conference in Leningrad again and since then one of the most frequently asked questions is whether one notices things have changed since the Berlin Wall came down. Are glasnost and perestroika visible? Indeed, I do think that is the case, but not in the Russian laboratories. During our four-day conference we had several concerts of religious Russian music and those of us who had already arrived in Leningrad before the weekend also attended the first public church service in sixty-two years, a religious service in the Isaac cathedral that was repeated on television every evening of that week. This was, we were told, because Boris Yeltsin and the mayor of Leningrad were sitting in the first row. On Tuesday evening the vespers of Rachmaninov were performed in the Smolny cathedral. Ever since Lenin had established his headquarters on the Smolny Island in the Neva most Leningrad inhabitants had not been closer to the golden dome than Oblomov and Olga Sergeyevna in their rowboat. The cathedral was packed with people who could not hold back their tears during the vespers. On Thursday evening the choir of the Physical-Technical Institute sang religious songs for us in the conference hall. The director told us that they had been invited to come and sing for the Romanov family last year. Then the KGB had forbidden them to sing the words – they were only allowed to hum the melody – but now at last it was possible to sing the religious text at the top of their voices.

Glasnost and perestroika are emotionally visible in Leningrad's churches, but not in the laboratories. In the Soviet Union scientists are still living an isolated life. Stalin forced every neighbour to be a spy. In Brezhnev's time someone you had never met before was a potential spy. This has led to unimaginable isolationism. The Cold War has isolated Russian scientists from the West, but what is worse: the Communist party, the system, has alienated Russian scientists from each other. That still has not changed and this became clear during our conference in Leningrad.

A meeting had been organized, sponsored by the European Physical Society, on the physics and technology of semiconductors, the material needed to make computer chips. From all the large laboratories of the Western European universities and from national and industrial laboratories representatives had come to Leningrad in order to meet their Russian peers for the first time. To our disappointment, however, the Russian representation consisted of physicists and engineers from the Academy only. Our host in Leningrad belonged to the Academy and that is why he had only invited people from the Academy. These people did not only come from the research laboratories, but also from the development departments and from the factories. Because the Academy has its own factories for scientific instruments, everything is developed and produced inhouse, from bolts and nuts to the micro-electronic chips they need. In the construction departments there are three hundred and in the workshops there are at least three thousand men and women working just for the Academy. Of course chips are also needed in computers for education, in hospital equipment, radio's, television sets, cars, telephones and for the military. Russia has separate ministries for science, for education, for healthcare, for radio, for TV, for cars, for telecommunication and for the military. They all have separate development and manufacturing facilities for computer chips. Everybody works entirely independently from each other. People in the Academy hardly know anybody outside the organization. One does not know the experts in one's own field, nor the experts in different areas. That is why the specialists outside the Academy had not been invited to our conference.

The isolation forced people to solve their own problems and they succeeded in doing so. Russian physicists and technicians are very well informed and can do everything by themselves. In Russia they have extremely capable professional people. In the laboratories of the Academy you may come across very ingenious equipment. In energy research centres plasma physics has reached the highest standards in the world. The space programme of the Soviet Union shows that also military laboratories are capable of outstanding achievements. And yet you will come across very frustrated colleagues everywhere. The Russians appreciate that in spite of their knowledge and skills they have lost out and the gap is still growing. This is not in the first place due to the Cold War and the economic boycott from the West. When the trade embargo will be removed, Russia will still not make a big jump in modern technology, for the Russians have fundamentally unlearned to co-operate within the Communist regime. The system has turned every colleague into a spy.

1990

B. Manfred Ullrich

In 1975 the president of the California Institute of Technology organized a ceremony when the news reached him that one of the Caltech professors was appointed a member of the Academy of Sciences of Romania. This inspired another Caltech professor, James Mayer, to get it into his head to found the 'Kaiserliche Königliche Böhmische Physical Society' together with colleagues at the Faculty of Electrical Engineering. Documents were printed on rice paper, decorated with a black eagle, a great red lacquer seal and the signatures of H.H Küllen (president) and B. Manfred Ullrich (secretary). Dozens of colleagues of Jim Mayer, spread all over the world, received such a document, together with an accompanying letter saying that he or she was appointed Member of the Böhmische Society because of the great accomplishments in ... (and then followed his or her speciality). Along with these documents letters were sent to the lab directors of these researchers or to the dean or rector of the university, which indeed had the expected effect. In several laboratories and universities all over the world colleagues of Jim Mayer were celebrated by their bosses. The story goes that at the IBM Research Centre in York town Heights work was terminated in order to be able to honour in public three co-workers for their breakthroughs in ... There is also a photograph of scientists at Bell Labs proudly showing their documents to their director.

In 1975 during the conference on Ion Beam Analysis in Karlsruhe, also the first Böhmische meeting was held. The members listened to a talk about 'Wine Analysis with Ion Beams'. After a short presentation the members of the Böhmische Society lifted their glasses. This made the other participants at the Karlsruhe conference jealous, not so much because of the wine but rather the exclusive character of this strange society. Professor Jim Mayer surrounded himself with a board of wise men, which became responsible for appointing new members. At present the membership is well above three hundred, most of them have been put forward by friends and colleagues, some have presented themselves as a potential member. One member of the board has taken the trouble of writing a lengthy letter to prevent one of his own colleagues from becoming a member of the Böhmische Society. In the meantime more than twenty Böhmische meetings have been held, in which talks have been given on a wide variety of subjects, such as the use of ion beams in archaeology, in astrophysics, in modern materials science, in the analysis of art objects, on particle accelerators and music, etc.

A few years ago a colleague from Uganda got into serious trouble at home. Apparently he fell out of grace with Idi Amin. The membership list of the Böhmische Society was used to urge people to write letters and send scientific publications to this colleague in Uganda, who was in danger of getting completely isolated. It was with great joy for all the members when at the following Böhmische meeting he showed up again. Also when the Chalk River group was afraid to lose their budget, the membership list of the Böhmische Society was used and letters of recommendation came in from all over the world. Therefore, the group could move to a new laboratory in London, Ontario. Not long ago, Professor Jim Mayer was called by a colleague who during his application for a new job was told he could not possibly be such an expert in the field he had mentioned as his name did not appear on the membership list of the Böhmische Society.

On the business card of Liu Bai Xin, following his name it says: Member of the Böhmische Physical Society. With a grin on his face he explains: 'I did not have anything else to put behind my name; since the Cultural Revolution the doctoral degree had been abolished.' Now Liu Bai Xin is Associate Professor in Engineering Physics at the Tsinghua University in Beijing. In 1981 he was one of the first to make use of the 'open policy' in his home country and he worked for a year at Caltech. Now he is one the few Chinese who is a member of the Kaiserliche Königliche Böhmische Physical Society; among a billion Chinese he has something that makes him special.

If you search the computer at your university library for the publications of B. Manfred Ullrich, you will get a long list with more than one hundred scientific publications on ion beams and their applications, all from after 1975. Following Jim Mayer many colleagues have taken up the habit of adding B. Manfred Ulrich's name as co-author on their publications. Most do this just for a joke, some because they want to show they belong to this prestigious society. A few chose B. Manfred Ullrich to have at least one co-author so that in the publication the word 'I' may be replaced by 'we'. B. Manfred Ullrich is not only a prolific writer, if you look into the Science Citation Index you will find that he has become also one of the most cited scientists of our time. Recently a rather thick volume appeared at Elsevier Science Publishers with only one author on the cover: B. Manfred Ullrich. It is the proceedings with three hundred contributions to a big international conference on the use of ion beams, organized by Professor Jim Mayer. It took him a lot of reasoning to persuade Elsevier to keep his own name from the cover. In the introduction to his book B. Manfred Ullrich explains that this book is the result of the joint efforts of all colleagues in the field, who all contributed as researchers, conferees, authors and referees, with the result that everybody should be proud of and for which B. Manfred Ullrich is grateful to everybody.

I wonder if the learned societies, such as the Royal Society in London or the Royal Netherlands Academy of Sciences were founded in a very different way. In any case, during my life as a scientist the Böhmische Society has been more than my peer group, even though we live half a globe apart. B. Manfred Ullrich is one of my very best friends who has made all the difference in my life.

1991

To colleagues and friends, for decades of Fun,
Utilization, Theories of everything and Survival,
Thank you

Diary of a fusionist

Friday 21 April. 'We have detected neutrons,' says Aart Kleyn, one of our group leaders, on the telephone, 'There were many. Much more than Fleishmann has reported. Our apparatus was running on deuterium for quite some time when suddenly the neutron counter went off scale, 300 counts in 3 seconds. Then I pushed the emergency button, but it was already 50 mRem, the maximum dose for a week.' We agree that I will look for more neutron counters, which we can read at a distance, before tomorrow morning when Aart and his group will be in the lab again.

This evening we listened to one of my PhD students who played cello in the Amsterdam Concertgebouw. It was marvellous. I still have my coat on and go to the lab to have a look. It is dead empty, as if the neutron bomb has gone off here. From the book at the reception it turns out that at least thirty people must have been inside the building when the cold-fusion experiment was being done.

In a rather original manner: in a vacuum chamber a titanium target is bombarded with a strong current of deuterium of 1 Amp at 100 Volt. This is the apparatus built by Ron van Os for conventional fusion research, on which he will graduate next month. Therefore the apparatus is available for something new and he makes use of it together with his successor. Are the ideas of Fleischmann, Ponse and Jones correct after all? I still cannot believe it, but the neutron counter says 371 indeed.

At our neighbours', the nuclear physicists, the experiments are in full swing and I ask for a neutron counter. 'For cold fusion?' They laugh at me. When I get into bed at 1:30 a.m., I hear: 'Did they measure cold fusion?' 'I don't believe it, but there are certainly neutrons.' I say and still try to get some sleep.

Saturday 22 April. At 1 p.m. the whole group is present. A provisional control desk is placed behind a concrete wall at about 10 metres from the apparatus. There they can read the neutron counters, except one on which they keep an eye through the video camera. There is one potentiometer with which the deuterium current can be adjusted, and the emergency button that will switch off the computer-controlled equipment. On the table is the science section of the Volkskrant (a Dutch newspaper) with a big article on cold fusion and the rather compassionate statement that in the Netherlands we still have not detected any neutrons. 'Wait for us,' Ron van Os says, 'in a quarter of an hour you will see.' While I try to get some work done in my room a colleague says that it is almost like a delivery. The newborn is announced, the mother is in labour but you cannot do anything but wait.

We have locked the lab. If you want to come in you have to call to the fusion experiment first. There a plotter draws a straight line on the paper to show there are no neutrons. In the library I find an old publication from 1948 in which fusion was reported by bombarding heavy water ice with deuterium. Below the acceleration threshold there is no signal at all and we are still a factor 100 lower in voltage. Right behind this article I see a publication by Edward Teller, the father of the hydrogen bomb, who calculated the threshold for fusion. Has he been wrong for 40 years? I cannot believe it, for all this research was done in the big bomb labs.

Our apparatus still has not detected anything, but I have to be patient, they say. Yesterday the meter went into the corner after the titanium was so overloaded with deuterium that it came out on the backside. I go home to see if I can help prepare the party of our youngest son. Everything is ready, the second floor is converted into a disco, and I quickly return to the lab. At the front door I suppress my inclination to ignore the barrier. I call the people at the apparatus and say: 'I can walk on through, can I?' 'Yes, but we have a signal!' Still not believing it, I join the group where the plotter now nervously moves about and slowly but steadily climbs. Every minute one of the co-workers walks around the concrete wall to read the third neutron counter. The video camera has stopped functioning. Jammed? Where we are the radiation level is zero, but at the apparatus not and at the neutron counter that is read every minute the signal rises rapidly. Soon it goes so fast that everybody gets excited and only Aart dares to step around the concrete wall to read the meter. When it is above 30,000 counts it is too much for him and he pushes the emergency button. Suddenly the apparatus stops and all the neutron counters fall silent.

After dinner I briefly meet with the group. We are still unhappy about the video camera being so sensitive to jamming whereas the monitor acts like normal and even shows Dutch TV programmes. Perhaps the jamming comes from our own equipment and it also affects the neutron counters? I say: 'If you do one more experiment with hydrogen and the neutron counters stay silent then you have one more proof than Fleishmann and Jones.'

Ron van Os does not agree. He argues that the deuterium discharge should produce a very different jamming signal than a hydrogen discharge in his apparatus.

I have to go home, before forty boys and girls will arrive for the party. Our family and friends will be there too, for my own belated birthday celebration. While upstairs the party is in full swing, I appreciate having made my best friends in our student days and not at high school. At about 11 p.m. I sneak out and call the lab. Aart Kleyn says he does not trust things any longer. They were able to get the video camera working again and they repeated the experiment. When the neutron counters picked up a signal the video camera went blank again. So, we probably pick up electronic noise and not neutrons. At 2 a.m., after the last guests have left our house, I do not have the energy to go into the lab again. The group will probably not be there anymore. It is only electronic noise and not neutrons.

Sunday 23 April. We went to the flower show in Limmen with all the foreign guests at our lab. The mosaics and the flower fields are always a great success. It must be wonderful to be a bulb grower and to open your curtains in the morning and all of a sudden see your fields not green but yellow and red all the way to the horizon. The landscape artist Christo cannot possibly compete with that. When I come home Aart Kleyn calls. They continued the experiments Saturday night until 3 a.m. and all the results reproduced beautifully. After that the deuterium was replaced by hydrogen and after more than an hour still not a single neutron was detected. Cold fusion looks OK. It remains strange how the video camera reacts during the experiments. Tomorrow they will continue.

Monday 24 April. Near the apparatus I find the measurements of Saturday night. The neutron signal from all three detectors nicely rises with the deuterium current. Above the graph it says in capital letters: 'CONTROLLED COLD FUSION'. I put these measurements on the copy machine, put them into my briefcase and go to the office of FOM in Utrecht. I have agreed to replace our deputy director in a meeting on personnel affairs. He is taking some rest after a jazz weekend in Schagen. The atmosphere during the meeting at FOM is very good. Science policy is first and foremost personnel policy and with these colleagues that is in competent hands. At the end of the meeting the measurements burn in my briefcase. I would love to say: 'Look here, guys, what we have: "controlled cold fusion."' I hold my breath and return to the lab for a work discussion with the cold fusionists.

On the blackboard as many as thirteen points are listed, things we still do not trust and which need further investigation. The most pressing is the video camera. Aart says: 'If it had not been there, we would now be drinking champagne.' The tasks are divided. We will rent video cameras, TV monitors and a diesel generator, so that all neutron counters and the video equipment may be noise-free and not coupled to the grid. I make sure that we get more deuterium, which I fetch from Leiden before the monthly meeting with my professor friends starts. The experiments can only begin at 11:30 a.m., until then we have the Dutch Society of Instrument Engineers visiting to teach them how we do computer-aided engineering. They are primarily interested in how we couple the computers from the design department to the computers in the workshop. After everybody has left the lab, the cold fusion experiment is started up again.

The neutron counters have been calibrated and we also know in the meantime that our video cameras are not sensitive to neutrons. Since the neutron counters get their power from the diesel generator outside the lab they are much quieter. Every hour I go and have a look, but the neutron detectors stay silent and the video cameras do not pick up any noise. Slowly it becomes clear that throughout the weekend we have only been measuring electronic noise and no neutrons whatsoever. At 3 a.m. we also know exactly where the electronic noise comes from and we close down.

It is 3:30 a.m. when I come home and hear: 'And is there cold fusion?' I answer: 'No, the video camera has saved us.'

1989

Roger

Suddenly the alarm went off. Everybody froze and looked at the red blinking lights above the gates. Two officers in white uniforms walked past the crowd towards Roger, who stood somewhat silly under the lights. With their detectors they skimmed his clothes and took him inside. The lights went off and at ease we shuffled through the gates. In the bus I heard this was not the first time Roger had made the alarm go off. The health surveyors already had taken him apart earlier. He had had to take his clothes off layer after layer, but no radiation was detected on his clothes or on his hands. Yet the alarm went off as soon as he stepped in front of the monitor. So he also had to take off his underwear and give it to them, but again no radiation was detected. When he stood naked in front of the health surveyors and they skimmed the detector over his body it turned out his penis was contaminated with radioactivity.

Roger worked in the plutonium laboratory at the institute where I worked as a young postdoc. He was world-famous for his knowledge and notorious for his negligence. At the end of the day before he left the lab he had not washed his hands and had not monitored them. Apparently he went to the toilet with radioactive contamination on his hands. After he urinated he did wash his hands, but in the meantime his penis was already contaminated. This goes to show that nuclear scientists must wash their hands before and after they go to the toilet.

1975

Solar cells

A young Dutch physicist presented two papers on Dutch solar cell research at the Twentieth Solar Cell Specialists Meeting in the American state of Nevada. Solar cells convert light directly into electricity, and the first paper showed that a formula has been found which makes it possible to calculate the electron current within the solar cell. The second contribution to the meeting demonstrated how the efficiency of solar cells from the Dutch company R&S was raised a full point to more than 10 per cent. However significant, these were only two contributions out of several hundred presented at this international conference. What is it that motivates the Dutch to be interested in solar cells? Wouldn't it be better to leave that to the researchers from 'the land of the rising sun'? Or to the Americans, who use more energy in summer than in winter because in summer they all have their air conditioners on?

On the question why the Dutch should be concerned with solar cells, I have eleven plus one answers:

1. In this country hardly any oil can be found, yet one of the largest oil companies is based here and Rotterdam is the largest oil harbour in the world.
2. In spite of the rain there is enough light in our country. There is only a factor 2 difference with the Sahara. If we would cover the roofs of our houses with solar panels, we would be able to supply the electricity needs of our households.
3. The price of solar energy is decreasing rapidly and the realization is growing that the social costs of fossil fuels and nuclear energy are high. The European Union has ordered a study of the environmental impact of energy consumption. If all costs were added to the retail price of electricity, it should be doubled. Wind energy would already be cheaper now and solar cells will be in ten years' time.
4. In West Germany, not an entirely sunny country either, the govern-

ment (under pressure from the Green party?) has decided to decrease the use of fossil fuel considerably in the coming fifteen years. To that end a hundred million Deutschmark per year have been put aside for research on and the development of solar cells.

5. Wind and solar energy can very well be put to use in combination with our electricity network, for 'peak shaving'. This should play a role in the decision to build new power plants, as a study at Utrecht University has shown.

6. Already now there are products, based on solar cells, in which Dutch companies are specialized and have a large advantage over the international competition. Take for instance the 'King of the Buoys', Stromag in Noordwijk (the Netherlands). This company places light buoys in waterways and harbours all over the world. There is a lot of interest for their most recent floating light buoy, which has batteries that do not have to be charged because this is done automatically by solar cells.

7. For space research Fokker (the Netherlands) is building satellites. These are entirely dependent on solar cells for their energy supplies. The efficiency must be as high as possible and perhaps that can be done with new kinds of cells developed at the Radboud University Nijmegen.

8. Philips Lighting is working together with R&S, solar cell producer within the Shell group, on lamps that do not have to be connected to the grid. So far they have developed garden lanterns with built-in batteries that are charged by solar cells in the armature. Not only garden lights but also all our electrical appliances are used only a few hours per day. In the remaining time they could be charged by solar cells off the grid.

9. The more the market for this kind of equipment expands, the more the price of solar cells will go down and that will create new possibilities for applications and new markets. In the near future billions of solar cells will be needed all over the world. If the Netherlands could supply only a few per cent of that market, it would be very attractive from both an economical and a technical point of view.

10. The largest markets for solar cells will be found in sunny countries without an extensive electricity network, such as countries in Africa and South America, China and India. These countries are too poor to import solar cells for remote villages. One wants to produce solar cells locally. That is not a problem, because it does not need any strategic materials. The most important element is know-how; which in the Netherlands is abundantly present. The solar cell market is an export

market, not of sun and cells but of technical knowledge.

11. The efficiency of solar cells in practice is not even half of what should be possible in theory. In countries with less sunshine people will be more motivated to go into greater depth with research and development, because the efficiency immediately translates into the number of cells needed for certain applications. Just look how thrifty we are with our sunny days.

A conclusive argument for solar cell research came from the young Dutch scientist himself, when he applied for a FOM position as a graduate student two years ago. During his interview it soon became clear that he was an exceptionally talented candidate, for whom no problem would be too difficult. I could not help asking why he did not apply for a different place in rather more fundamental research in physics, instead of this more engineering work. His answer was that research costs a lot, primarily from the researcher himself, who is usually totally absorbed by the research. It is quite often so absorbing that it takes up all his attention and leaves no time, nor interests, for anything else. For solar cells, however, this young physicist found the research effort worthwhile.

1988

Particle accelerators

There are two different kinds of physicists who use particle accelerators for their research. One group is searching for elementary particles. The other group uses particle accelerators to make and analyze new materials on an atomic scale. The elementary particle physicists are building increasingly larger accelerators in order to reach deeper into the atomic nucleus. The materials scientists use the 'leftovers', the accelerators that have been used to split atoms but which are no longer of interest for nuclear physicists. It is interesting that industry has kept in pace with these developments. HVEE in Amersfoort, the same company that used to build particle accelerators for the nuclear physicists, now also is building accelerators for materials scientists.

Elementary particle physics has gradually been concentrated around gigantic accelerators only in a few research centres in the world. Over the last forty years some thousand small accelerators have been installed in universities, governmental laboratories, and in industries to facilitate materials science. As far as the accelerators of CERN in Geneva are concerned, it is common knowledge that they are being used in the search for the quark, gluon or Higgs-boson. What are all the other accelerators for?

By far the largest number of accelerators may not be found in science labs but, you will never guess, in microelectronic factories in order to make computer chips. More than any other industry the microelectronic world is the pioneer of modern technology. Recently there has been some negative publicity surrounding Philips and other chip producers, but there is no section in industry that spends such a large amount of money on research and development. In ten years' time the electronic industry has grown to the size of the car industry, but in microelectronics a lot more is done in research and development. This has led to an improvement of the price/performance ratio that has never been seen before. I have heard the following comparison being made: what has happened in microelectronics is like making it possible to buy a Rolls-Royce quality car for the price of a Matchbox! This is mainly due to modern materials science. We have learned to build complex electronic circuits and memories on the surface of a silicon wafer, using modern physical methods such as ion implantation.

The particle accelerators that were used earlier for nuclear physics, nowadays serve as a cannon to shoot atoms into the surface of the chip, exactly where they are needed to change the electronic properties of the chip material. In this way certain locations in a silicon wafer are made conductive and other places isolating by implanting foreign atoms with more or with fewer electrons than silicon. Particle accelerators are used for this because nuclear physicists have shown that it is possible to shoot with only one specific kind of atom whereas unwanted atoms do not penetrate the material. Due to the large accuracy, reproducibility and uniformity, the technique of ion implantation has become popular in producing chips to such a degree that only for this application three thousand dedicated accelerators are spread all over the world.

In other areas the use of particle accelerators is still in the stage of research and development. That is possibly even more interesting because new materials properties are still being found. For instance, it has been discovered that metal surfaces become hard as stone when they are bombarded with nitrogen atoms. A protecting layer is formed which is resistant against wear and corrosion and often gives less friction. This may be used in ball bearings. In England the wheels of Formula I racing cars are mounted on ultra thin bearings. In order to protect the bearings they are implanted with nitrogen. Even though the surface layer is ultra thin and wears away in one race, the bearings are strong enough for a complete race, which would be impossible without the nitrogen ion implantation.

Corrosion protection is also important for artificial limbs such as knee and hip joints, which are often made of light metal. The surface of these artificial limbs is implanted with nitrogen ions before they are put in place; this will reduce the influence of body fluids and prolong the functioning.

Also in the chemical sector particle accelerators are beginning to play a role. In Italy it has been discovered that pretty much every materials property can be synthesized with the use of ion implantation. Polymers lose hydrogen when being hit and the remaining carbon forms a layer that hardly differs from diamond. This extremely thin layer of diamond-like material is invisible, but isolates well and is extremely hard. If the bombardment takes too long the thin layer loses all its hydrogen and a disorderly layer of graphite results. This is a conducting layer much like metal. So, in principle it is possible to make surfaces conducting, semi-conducting or insulating, hard or soft, transparent or black.

Also for technical ceramics interesting effects of ion implantation are reported. These materials are known for their hardness. After implantation the hardness may still increase whereas the fragility decreases. The extra atoms, after having been implanted into the surface, provide a tension that holds possible breaking crevices together.

In glass the index of refraction is totally changed by ion implantation. This is used to make light conductors, switches, amplifiers and even lasers on a single piece of glass. This may become important for telecommunication purposes and might lead to superfast computers working with light.

Up till now it was only a matter of implantation into dead materials, but particle accelerators are more and more being directed to living matter also. From China it has been reported that rice seeds have been implanted with the result that the harvest has definitely improved.

In some accelerator labs positive experiences have been obtained with treating tumours in animals, thus preparations are being made to treat people as well. The big advantage of ion beams over the use of x-rays is that they deposit their energy mainly at a specific depth, and might be able to kill a tumour there without causing too much damage to the healthy tissue around it. Sadly enough it is only possible to use this method for shallow irradiations, because the particle accelerators that are available for this do not have such a high energy, and the ions do not penetrate so deep.

Not one nuclear physicist has been able to foresee that particle accelerator developments would go in this direction. In the second half of the twentieth century nuclear physicists built more and more powerful machines, and the accelerators they left behind often came into different hands and were used for unexpected applications both in making and in analyzing modern materials on an atomic scale. Despite this success it is not to be expected that particle accelerators will continue to have the same effect in the coming years. The size of the largest accelerator at CERN in Geneva and the budget and time needed to build it are so phenomenal, it is likely that the physicists have reached their limit. Moreover, I think that the twentieth century was the century of the physicists, whereas the twenty-first will be the century of the biologists.

1990

The Silicon Age

After the Stone, Bronze and Iron Age we now live in the Silicon Age. Our daily life has been changed by a little chip of silicon that takes over dumb jobs and creates new work. In physics research the chip has made a revolution too. Silicon in itself has become one of the most important subjects of research. Experiments without electronics and information technology are unthinkable in these modern times and computer simulations are replacing traditional 'Gedankenexperimenten'. This I will illustrate with examples from fundamental research on matter; three different kinds of investigations: 1. for computers, 2. with computers, 3. within computers. After that I will describe how changes in the Silicon Age lead to significant differences in scientific research in Russia, the US, Japan and in our own country, differences originating from discrepancies in the local economy, politics and the military.

1. For computers

Oil may be simply pumped up, but the computer, the TV and the telephone result from much research and development. The biggest and best research laboratories of the world were founded by AT&T, IBM, Philips, Hitachi and other electronic industries. They employ many thousands of scientists and engineers working on the next generation of chips, lasers, detectors and actuators, which are applied in consumer electronics, in cars, in professional instruments, in rockets and other military equipment: 'Big science for big business'.

Modern microelectronics, chips, for a large part consist of the material silicon. No wonder that today more scientific papers and patents appear on this than on any other material. A whole new field of research has been created: the physics and chemistry of solid-state surfaces. Complete electronic circuits and memories are built in and on the surface of a silicon wafer. To this end foreign atoms are implanted into the silicon surface, which alter the electric properties of the surface and make certain spots conducting or just insulating. This requires knowing exactly which atoms are present in the surface and how they are oriented relative to each other. It turns out that the same physical measuring methods that were developed to explore the inside of matter are also apt to study the physics and chemistry of silicon surfaces. Using the equipment and the methods from atomic and nuclear physics one can also map the structure and composition of surfaces on an atomic scale.

The same particle accelerators that were developed after WW II by the nuclear physicists for the study of the atomic nucleus are used today to make chips. With these particle accelerators one can implant the proper atomic nuclei into the silicon wafer, exactly on the spot where they are needed to obtain the special electric properties. Thus we have learned to manipulate atoms with unsurpassed control and precision. Not only particle accelerators but also lasers, plasmas, molecular beams and electron microscopes have become standard tools in the workshop of the modern microelectronic world. With these many millions of transistors are fabricated on a few square millimetres of silicon. The electronics industry is insatiable and demands from the researchers require ever more complex circuits to be put on a silicon wafer. Thus nano technology came about, enabling to build the next generation of computers.

Not only computers are made using this new fabrication method, also other products profit from it. Thanks to nano technology the amount of information that may be stored on CDs and magnetic disks is enlarged, the resolution of video cameras and TVs is much improved; flat television screens and new telecommunication equipment have been made possible. Moreover, not only microelectronics benefits from developments in surface science, also the chemical industry can profit from a better understanding of chemical reactions on a surface. This is important, for instance, for catalysis. Surface treatment can also increase hardness of metal surfaces and improve their resistance against corrosion and wear. Therefore the physics and chemistry of surfaces also produce new products for the metal and chemical industry.

Nano technology has also become a discipline in itself and physicists and chemists have wondered: What is the limit? How far can they go on miniaturizing? In fundamental research they are already doing experiments with one atom, one molecule or a single electron or photon, one particle of light. This is possible thanks to developments in microelectronics and the surprising atomic scale control in the fabrication of new instruments and materials. Here we also reach a fundamental limit dictated by the uncertainty principle of the quantum world. Today one makes switches that are so small they only allow the transfer of one electron at a time. In principle such a switch should consume extremely little power and be very fast. According to quantum mechanics, however, it is impossible to predict exactly when the switch will actually function. We can open the gate for the electron and measure how long it takes the particle before it goes through. We can also switch a large number of times and determine the mean switching time, but for every individual electron we will never be able to predict how long it will take before it will go through the gate. It makes quantum electronics unreliable. That is why for the time being it will remain an interesting subject for fundamental research, but it is questionable if it will ever lead to applications of quantum switches.

2. With computers

Not only the receptionist and the management assistant, but also the scientist spends a considerable part of the day in front of the computer screen. Modern experiments are computer controlled. The computer does the measurements, following a programme that the researcher has developed. The computer stores the data in its memory, for further processing after the experiment is finished. This happens mainly because the computer can do much more simultaneously and more accurately than the researcher. The computer makes experiments possible that would otherwise be totally out of the question.

An example is the scanning tunnelling microscope that can image the structure of a surface on an atomic scale using the computer. Looking at the computer screen you think you see rows of individual atoms. Yet this is not for real, the image is generated in an ingenious way by the computer. The microscope scans the surface with a fine needle, just like in an old-fashioned gramophone, and measures the 'tunnel current' between the needle and the surface. What that 'tunnel current' is, we do not know exactly. It is a very small current of electrons that jump from the surface to the needle whenever the latter happens to be just above an atom. That is really very strange, for the electrons belong inside the surface. If they could just move out of it, the surface did not exist. But if the needle is right above an atom, apparently the electron feels the needle and has the possibility to jump over, to 'tunnel' through. We can only explain this using quantum mechanics, by assuming that the electron behaves like a wave with a wavelength long enough to stick out of the surface a little. As soon as the needle approaches the surface with a distance smaller than the wavelength of the electron, then the electron feels the needle and can 'tunnel' to it. It is as if there is a short tunnel between the surface and the needle, through which the electron can go. Like a lorry through a tunnel in the Alps can go from one valley to the next. Of course this is only an image, for the tunnel is not real, also not for the electron. We cannot see what happens on the atomic scale. Yet the computer allows us to use the process of tunnelling to image in full colour the atoms in the surface.

The distance between the needle and the surface we cannot measure, it is entirely controlled by the computer, which also measures the electron current. Using these data the computer produces an image on the screen. It is the tunnel current, which we see, as a function of the position of the needle above the surface, but it looks as if we see individual atoms. That is because the computer screen gives a picture that resembles the image we have of atoms and how they are ordered in the surface of a crystal. Because our mind's eye and the computer image are so much alike, we recognize it, even though we do not really understand what the computer shows. The danger is that we fool ourselves, and others, with these images.

The computer is not only a new tool allowing us to do experiments that were impossible before, but the computer also redirects our research interests. The computer and the microelectronics, with which the scanning tunnel microscope is connected, can process the data so fast that one makes ten images per second of a piece of a crystal surface. Thus video images are possible of the movements of atoms over the surface. Through these video images the researchers are fascinated by the dynamics of the surface, the motion of atoms, rather than the static structural phenomena. The computer creates new possibilities for experiments, this makes such experiments fashionable and they will replace other scientific questions for which computers may not be strictly necessary.

3. Within computers

One can also experiment using the computer without being connected to other measuring equipment. The computer stands alone, without any other instruments, and the researcher uses his keyboard to control a programme that runs inside the computer. In this way one can do research within the computer.

There are two kinds of research within the computer. One can calculate the properties of new materials by searching for data of known materials properties, combining them and calculating them further. In this way one can calculate, for instance, the structure of polymers and proteins before even making them. One can calculate how different materials will behave under extreme circumstances as well. One may calculate if certain alloys will be stable at high temperatures, high pressures, or under radioactive irradiation. One can test materials under dangerous conditions without having to simulate them in the laboratory. Only after one has learned from the computer simulations how one should work, the real test is performed.

There is a second kind of computer simulation. Here it concerns testing a theory inside the computer and it is best compared to a so-called 'Gedanken-experiment', which has been popular in physics, for instance in the discussions between Bohr and Einstein on quantum mechanics. One thinks of an experiment and tries on purely theoretical grounds to predict the outcome. As long as such experiments are simple we can simulate them inside our brains. More complex 'Gedankenexperimenten' require a computer to simulate them. This also offers the interesting possibility to simulate conditions within the computer that do not exist in nature but which may still give us new insights. In this way some noteworthy discoveries have been made of which I will give an example.

It concerns liquid crystals. They are well known, for they are used in displays of watches, pocket calculators and flat panel computer and TV screens. Liquid crystals consist of elongated molecules in solution. The molecular sticks move about randomly but they may be oriented in the liquid, put all in the same direction. It is this ordering which is used to make images with liquid crystals. Chemists wondered if ordering would be created purely because the molecules block each other's way, they repel each other. Some chemists thought that repulsion alone would not be sufficient and ordering of the molecules in the liquid required not only repulsion but also attraction. At large distances one usually finds attraction. Now from real molecules you cannot easily switch-off their attractive force, but inside the computer you can just do it. Using computer simulations of the behaviour of elongated sticks it is now well established that molecules do not need to attract each other to create order in solution. They show collective behaviour by merely repelling one another. Order from repulsion was discovered thanks to scientific research within the computer.

Although changes in science due to the computer chip are taking place all over the world, there are significant differences in different parts of the globe, which are caused by local economic, political and military factors. Research for, with and within computers is already dominating in physics laboratories all over the world, except in Russia. Until about ten years ago the former Soviet Union was a superpower, also in science. Through the boycott by the West, especially of computers, Russian scientists lost their leading position. Their scientific publications are no longer read, because they do not deal with the subjects and materials we in the West are busy with. They just do not have the opportunities. Their experimental facilities are hopelessly old-fashioned for they are still not computer controlled. Their theory lags behind due to the absence of numerical methods worked out by computers. One hardly hears anything from famous national research laboratories as the Ioffe Institute in St. Petersburg or the Lebedev Institute in Moscow. Thousands of scientists who used to work there are frustrated, not only because with their low salaries they live on or below the poverty line, but especially because they are no longer taken seriously compared to colleagues in the West.

The stage is set by the US, also the stage of the Silicon Age. Since the discovery of the transistor in AT&T Bell Labs, shortly after WW II, this laboratory along with the research centres of IBM and the National Labs have belonged to the most productive in the world. Together and in healthy competition these laboratories, and the American universities, have initiated research for, with and within computers. However, one does not always appreciate how much the research direction was dominated by military subsidies with which these developments were funded.

During WW II American scientists were extremely effective in developing the first nuclear weapons in the Manhattan Project. This gave politicians the feeling that it was worthwhile to keep physicists together and happy. Thus large defence budgets were created for research, with which the scientists developed new weapons, but they also were able to do fundamental scientific work. This they did so successfully that they could convince the military of the necessity to provide even bigger budgets for still larger research and development projects. The most recent programme is SDI, the costly shield in space that is supposed to protect America against inter-ballistic missiles. On this promise advanced research of the last decade has been financed. Due to its success physics in America, at universities, national labs and in industry has become dependent on defence budgets.

This has estranged American physicists from society. One does research primarily on subjects that have emerged from pure scientific interests and that may be sold to the military as relevant for the future 'electronic battlefield'. Thus supercomputers and advanced networks have been developed, instead of personal computers and microelectronics for consumer products. Superintense laser systems for the military have been made, instead of small lasers for CD players. New particle accelerators have been developed which could provide satellites with a 'death-ray', but which are primarily used in fundamental research.

The Cold War was won and undoubtedly the close collaboration between science and defence contributed to the victory. But American physics is in a crisis. Because there is no enemy left and because of the enormous budget deficit of the government, the defence budgets are reduced. Now the university professors have lost their research funds. The national labs have lost their reason for existence. And industry was used to fund risky research and development projects through the defence contracts, in name of national security. Government support for industrial products is taboo in America. It is seen as contradictory to 'the invisible hand' guiding the moral values of the free market and free enterprise. Consequently many researchers lost their job and America has lost its leading position in microelectronics to Japan.

On the list of consumer electronics Matsushita and Sony are the number ones, then Philips from the Netherlands and Thomson from France, then Hitachi, Pioneer, Toshiba and other Japanese companies. In the top ten there is no American company. In microelectronics more is invested in Japan than in America and much more than in Europe. In Japan part of the investments are paid by the Ministry for Trade and Industry (MITI). The remaining budgets have to be financed from the profits. We consumers are not willing to pay large sums, so the margins are small. In Japan there is hardly a defence budget from which developments may be paid. Therefore all the money that industry earns is spent on developments of new products and little money is left for research. Basic fundamental research of any stature is hard to find in Japan, there is no budget, nor tradition. This explains why Japanese physics is so mediocre even though Japan is the market leader in consumer electronics. Yet, not much new will emerge from this country as long as there is no innovative research.

In Europe the Netherlands is the only country that is not yet fully dependent on imports for the making of chips. ASMI in Bilthoven and ASML in Veldhoven make equipment for chip manufacturing and export these all over the world, including America and Japan. And Philips is still a giant in all sectors of microelectronics. This is possible only thanks to large numbers of well-educated people from which these companies can choose. People trained in electrotechnology, semiconductor physics and computer science, at Dutch universities.

In our country computer science draws large budgets and talents, much more than any other sector. But our society has many more priorities. In order to survive in the metal industry products with a higher added value are now already necessary. This means more research and development in, for instance, micromechanics. In our country there are multinational companies in chemistry. It would be desirable for these industries to be just as innovative as microelectronics. But chemistry is unpopular in our society. From microelectronics one may learn how many talents are mobilized for radical innovations. Or is this not to be expected from traditional chemistry and should we hope for developments in molecular biology and growth in biotechnology companies? There are still other priorities in our society such as energy and environment, in which much more innovation is possible, which should be stimulated.

The economic situation makes it necessary for Dutch companies to concentrate on their core activities. In industry laboratories for research and development are focussed on problems of today and new products for tomorrow. The research and development work is financially more and more dependent on and catering to the business results. Consequently, only few industrial research laboratories can afford to do fundamental research and long-term developments are in danger of being ignored. There is the tendency to help industry by doing research and development at university for the market, but that can better be done in the industry. It is the task of the government to create the circumstances necessary for our country to keep its top position in fundamental research; universities in which the best scientists and engineers are trained before they move on to society. This is the time to stimulate innovative research such that our country may hold its leading position, also in the Silicon Age.

1993

The discoverer and the inventor

Soon after Professor W.C. Röntgen had made his discovery public there was a visitor, Dr Max Levi, sent by AEG management to the University of Würzburg for an agreement to collaborate with the professor. Röntgen politely listened, he was duly impressed by the interest from such a large company, but he declined the offer. In his opinion scientific discoveries were to the benefit of mankind and should not be exploited by individuals through patents and licenses. On the opposite side of the ocean Edison thought very differently: 'Professor Röntgen probably will not make any money with his discovery. He belongs to those pure scientists who merely love to dive into their subject and uncover the secrets of nature. After they have discovered something wonderful, someone else has to come to look at it with a practical eye. That is what should happen with Röntgen's discovery. One has to look for practical value and financial revenue.'

In December 1895, before he had published his discovery, Röntgen made the world famous picture showing the bones in the hand of his wife with the wedding ring. He welcomed the excitement around the 'x-rays', which occurred worldwide within half a year (a hype in times without internet and when the telephone had only just been introduced), as a diversion from his research. Röntgen was more fascinated by the question of the true nature of the new radiation as is clear from the final sentence of his publication in which he does not write about possible applications, but rather: 'Sollten nun die neuen Strahlen nicht longitudinale Schwingungen im Ether zuzuschreiben sein?' Also the publications that followed dealt more with the origin of the new radiation than its applications. The inventor Edison, however, developed a fluoroscope for the electrical exhibition in New York on August 1896, so within half a year people could observe the bones in their own hands. With this machine Edison drew the largest crowd at the exhibition.

Today the Professor Röntgens are rare in the temple of science. Patents are fashionable. The craziest example comes from the geneticists who even tried to patent the human genome. Bill Clinton and Tony Blair rightfully barred that; one cannot patent 'the language of the creator'! Nowadays, we do not seem to distinguish between the discoverer and the inventor anymore, whereas the first discovers what was there already but what was never seen by humans before (such as x-rays), the second invents a useful application (to take an x-ray of someone). The discoverer publishes, the inventor patents, both sometimes like mad.

Some people think they are smart by keeping their work secret. They may be disappointed, for usually if something is discovered or invented the time is ripe and one may cry 'Eureka' in many different places. Those who keep their discovery secret, run the risk that a colleague is honoured instead. That was true also for x-rays at the end of the nineteenth century. Even though Röntgen received the Nobel Prize for it, in most countries one speaks of x-rays rather than Röntgen-rays. Scientific results are published in the interest of society, and that of the scientist, we add today.

The discoverer who keeps fundamental knowledge secret runs the risk of missing the Nobel Prize, the inventor who keeps useful knowledge to himself delays the application. It may sound paradoxical, but if you know something useful it is in the interest of the application that you should protect it with a patent. Thus one has the opportunity to invest in the invention, money that may be earned back by the revenues from the monopoly position obtained temporarily by patenting. But the inventor is not always like Edison, the most apt to bring the invention to the market. Through a patent the useful knowledge gains value and may be traded.

Today the discoverer and the inventor are stimulated to collaborate. Just as with x-rays most discoveries in the end turn out to be of some use. So it is worthwhile if the discoverer is stimulated to think about possible applications of his discovery and not only to publish about them but also to file a patent. If the Röntgens sell their patents to the Edisons, it is in the interest of science and its applications. Science is for society, not for the shelves.

1995

Jaap was right

The last time I met Jaap Rodenburg was at a reception in the queue of people waiting to congratulate the new Professor of Wind Energy at the Technical University Delft. In the reception book Jaap wrote 'Greenpeace' below his name, therefore I wrote 'Energy-research Centre the Netherlands (ECN)' below my name. I liked the look of that so brotherly together on the same page.

At meetings of the energy world Jaap and I looked for each other to argue about the same subject all the time. Jaap always wanted to be right, me too of course, and I most preferred to be granted right by him. This had been so since the hearing in Parliament on the reprocessing of nuclear fuel from the Netherlands. There I defended the position that reprocessing was best, but Jaap stole the show. He did not speak personally but gave his time to people who lived next to La Haque, Sellafield and Dounray, people who told terrible stories of the consequences of leaks and contaminations caused by these reprocessing plants. Especially Dounray was blamed and I took it personally as the ECN has a reprocessing contract with them in connection with the production of Molybdenum 99, a radioactive isotope with which at least five million patients in European hospitals are diagnosed each year. Mo99 is a fission product produced in the nuclear reactor in Petten (the Netherlands) that is separated from other fission products via chemical processing, after which the irradiated uranium targets are sent to Dounray for reprocessing. A nice example of recycling and an important medical application, what could Greenpeace have against that?

Jaap Rodenburg's criticism was heavy and well presented, with a great feel for drama. One citizen of Dounray came to tell us that the reprocessing plant was an old-fashioned mess. There was a serious shortage of trained personnel, they did not live up to the safety rules, at some places they had dumped so much radioactive waste that it had become critical and had exploded, and a castle in the neighbourhood, once a tourist attraction, is now so contaminated that it is closed to the public.

'How can you do business with such a company?' one Member of Parliament asked me. I answered that Dounray was part of the UKAEA and stood under international surveillance by the IAEA in Vienna and as long as Dounray had a license I had no reason to doubt the quality of their reprocessing facility, and that recycling of radioactive fission material should be better for the environment than using new Uranium all the time. Moreover, stopping the production of radio-isotopes would have serious consequences for at least five million patients in Europe alone.

Apparently the minister was convinced. But not Jaap Rodenburg and that I did not understand. Until last month when the news came: the UKAEA had ordered an audit at Dounray. The official findings were that they do not live up to the safety rules and that they cannot possibly fulfil their license, which as a result has been taken away.

Jaap was right, but now I cannot grant him this because Jaap is no more. In the reception book of the professor in Delft my name is written below Jaap Rodenburg, Greenpeace.

PS

In the summer of 2001 I discovered that our nuclear research and consultancy group in Petten had secretly initiated the building of their own reprocessing plant, because the highly enriched uranium remnants from the Mo99 production could not be sent to Dounray anymore. Issues of environmental impact and the non-proliferation treaty were waved aside by referring to the medical applications. I could block this development in time, but only after seeking the support from ECN's Supervisory Board.

On a winter night in December 2001 there was a power failure in North Holland, where Petten is located. The nuclear reactor is a research reactor, not a power reactor; it needs electricity to operate, for instance to pump cooling water. The reactor has a back-up cooling system to prevent meltdown of the core in case of a power failure. But this evening the back-up cooling system failed to come into action and the operators did not know what to do. There is an extra safety system by convection cooling for which the operators had to open a valve, but the control room was dark. When they reached for a torch that should have been there, it had been taken away by a colleague to work under his car. Trying their luck the operators put the valve of the convection cooling in what they thought was the 'open' position. But then the lights came back on and the operators discovered they had actually closed the back-up convection cooling system. Had the power failure lasted longer it would have meant meltdown and a major disaster. When I learned about this some months later – they thought they could keep it secret – I did not think I could take responsibility any longer and I resigned from the ECN.

Moore's law in Bilthoven

When the new annual reports arrive again, the ones from last year go into the recycling bin, except for the reports of ASM International at Bilthoven. Those I have kept since 1981, when I became advisor and later board member. In 1981, their revenue was 100 million guilders and the profit seven and a half million. Then ASM shares brought in more than 20 million dollars at the NASDAQ stock exchange in New York. In the annual report of 1999 ASM International reports total sales of 414 million Euros and eleven million profit. In the meantime, the market capitalization is one and a half billion dollars. That is the effect of Moore's law in Bilthoven.

After the discovery of the computer chip in 1961, Gordon Moore set himself the goal of doubling the number of transistors on the chip every year and a half. Because his company, the American firm Intel, has succeeded to meet this goal for almost forty years, Moore's law has become a standard in the computer industry. Consequently, since 1961 the cost of a transistor has been reduced by six orders of magnitude. Every new generation of computer chips is ready after one year and a half, costs the same as its predecessor, but has twice as much computer power. Therefore chips penetrate more and more into our society, first in the computer, then in the TV and now in the mobile phone and other consumer products. If you use a computer with a Pentium inside, that Pentium chip is made by Intel, but in order to produce that chip it has to go through equipment of ASM International. Thus Moore's law also holds in Bilthoven.

It has been a hard lesson. The annual reports from the 1980s still tell of ASM's ambition to deliver turnkey factories for worldwide chip production. Investments are made in Phoenix, in Boston. ASM is the first company from the Netherlands with its own production facilities in Japan. ASM's subsidiary in Hong Kong built two factories, one in Singapore and one in China. Of course the most important investments are made in our own country: in Bilthoven investments are made in an R&D centre and in Veldhoven ASM is starting a joint venture with Philips (ASM-Lytho).

Then something goes wrong with the computer industry. There is an overcapacity in the production of chips, the prices go down and one suffers losses. Only the financially strong remain. Fortunately ASM has built up a considerable equity, but the recession lasts and the research and development programmes take much longer than expected. ASM knocks at the door of the banks in vain, so company assets have to be sold. But how do you sell technology that is not yet ready for the market? Fortunately, the American firm Varian pays a nice sum for the Boston activities, but ASM-Lytho must go back to Philips at a loss. The R&D centre in Bilthoven is closed down. Reorganization is announced in Phoenix and in Japan. After thorough analysis a beautiful factory in Gelderland changes hands. Almost half the shares of the company in Hong Kong are taken to the stock exchange there. This is painful for it means financing losses. Finally only three strategic techniques remain in the hands of ASM International. Fortunately they turn out to be winners. Developments in Phoenix are so successful that ASM becomes market leader in epitaxy equipment. Unfortunately the patent portfolio is not guarded well enough and the competition starts a lawsuit without any ground, but it costs ASM buckets of money. In the end as much as 80 million dollars have to be paid to get rid of the awful American lawyers. ASM survives this bloodshed and the production is moved from Phoenix to the Netherlands.

If you read stories like this in the paper you do not always appreciate what it means to live in a world ruled by Moore's law. The ambitious growth of our economy always generates overcapacity, which all the time claims its victims. ASM also has learned its lesson as is witnessed in the annual report of 1999. Not much is left of the ambitions of the 1980s. We have lost an illusion but in its place we have gained some carefully selected innovations. These technological highlights are so outstanding not a single chip manufacturer can do without, not even the maker of the Pentium. Therefore, it is a shame that one typical element in the 1999 report is missing: the secret of ASM in Bilthoven, through which it has survived. In the 1980s the reports always showed a picture of Arthur del Prado, the founder and boss of ASM International. His friendly face looks at you, with an open expression that tells you exactly what he is up to, you see the determination in his eyes: Moore's law may be the rule of the industry, in Bilthoven Arthur's law is first to survive.

1999

Teller in the Netherlands

President Reagan was standing next to Premier Gorbachev and introduced me to him, saying: 'This is Dr Teller'. I put my hand out to shake hands, but Gorbachev stood unmoving and silent. Reagan then repeated to Gorbachev: 'This is the famous Dr Teller.' Gorbachev then said, with his hands at his sides: 'There are many Dr Tellers.'

If the bombs on Hiroshima and Nagasaki in August 1945 had not forced the Japanese to give up, then my wife and mother-in-law would not have survived the Japanese camp. Yet I am ashamed to have shaken hands with the man who has devoted almost his entire life to the development of weapons of mass destruction. It was 1975, the Cold War was at its peak but also the anti-neutron bomb demonstrations, when Edward Teller came to visit our lab in Amsterdam. My boss, Jaap Kistemaker, not afraid of anybody or anything, received Teller in his office where I, being a group leader, could not be absent, but I did not attend Teller's colloquium.

That afternoon we heard that a demonstration against our lab was to be expected and that perhaps they would occupy the building. Jaap Kistemaker requested us to stay in the lab overnight and from stores he brought long metal bars, which he gave us 'for defence'. That evening a few activists indeed have demonstrated in front of our lab, but they did not stay. They merely shouted 'Kistemaker A-bomb maker'.

Whenever we went to my mother-in-law for the weekend and the train passed through the Zaan, the region north of Amsterdam and stronghold of the Communist Party, I would always see that same text in big white crying letters on the fences of the companies: 'Kistemaker A-bomb maker'. It made me uncomfortable – I knew it was a lie, in our lab it did not even come close to working on the A-bomb. Also not in the early days, when the ultracentrifuge was developed for enrichment of uranium. This was merely for low enrichment levels, not enough for bombs but only for nuclear energy. 'Atoms for Peace', President Eisenhower's predicament was also Jaap Kistemaker's conviction, but the general public did not believe it, and the visit by Teller, although it was short, of course did not help.

It was good that I did not attend Teller's colloquium, otherwise I would certainly have asked him the following question: 'Can you tell us, Dr Teller, how come you are seen as the real Dr Strangelove, whereas it was just announced that Andrei Sakharov will win the Nobel Peace Prize?' Indeed, to my great surprise in that same year of 1975, the father of the Russian H-bomb was awarded of all prizes the Nobel Peace Prize. I have always wondered why the one was despised whereas the other was declared saint. How was this possible? I will try to answer that question here.

In his memoirs Teller writes about the first test of the atomic bomb:

'We all were lying on the ground, supposedly with our backs turned to the explosion. But I had decided to disobey that instruction and instead looked straight at the bomb. I was wearing the welder's glasses that we had been given so that the light from the bomb would not damage our eyes. But because I wanted to face the explosion, I had decided to add some extra protection. I put on dark glasses under the welder's glasses, rubbed some ointment on my face to prevent sunburn from the radiation, and pulled on thick gloves to press the welding glasses to my face to prevent light from entering at the sides.

For the last five seconds, we all lay there, quietly waiting for what seemed an eternity, wondering whether the bomb had failed or had been delayed once again. Then at last I saw a faint point of light that appeared to divide into three horizontal points. (It actually was the nuclear explosion and the luminous ring around it.) As the question "Is that all?" flashed through my mind, I remembered my extra protection. As the luminous points faded, I lifted my right hand to admit a little under the welder's glasses. It was as if I had pulled open the curtain in a dark room and broad daylight streamed in. I was impressed.

A few seconds later we were all standing, gazing open-mouthed at the brilliance.' (*Memoirs, a twentieth-century journey in science and politics*, Edward Teller with Judith Shoolery)

Just before the test was executed, Fermi asked the question what the chances were that the nuclear explosion would start a chain reaction such that the whole atmosphere would explode. Apparently nobody had yet thought of this possibility. Teller was put on to this problem and produced a memo with possibilities and impossibilities. On the way to the first test he was still discussing it with one of his colleagues who asked him what he would do in case one of the possibilities would turn out to be true. The just as happy as irresponsible answer was that he would swallow a second bottle of whisky.

Although most physicists who had worked in Los Alamos on the atomic bomb, after WW II looked back with some feelings of guilt over the devastation at Hiroshima and Nagasaki, for Teller it was by far not enough. In his memoirs he explains extensively that fission bombs, such as were dropped on the Japanese cities, could for physical reasons hardly be made bigger and more powerful. Whereas fusion bombs, such as the Hydrogen bomb, have no limits and thus for strategic reasons they are much more attractive. While Fermi and Rabi warned for the consequences of such enormous explosions on the atmosphere, Teller continued saying that his estimates of the effects of enormous explosions on the atmosphere were limited. Teller gambled. Only years after the first test explosions of hydrogen bombs computers became large enough to simulate the effects of enormous explosions in sufficient details.

These examples from Teller's autobiography remain mind-boggling, even though we knew it from Teller all along. Reading Andrei Sakharov's memoirs, one discovers it was not very different with him.

'In fact I had no choice, but it was out of my own free will that I worked extremely hard and was totally dedicated to my work. Now, some forty years later, I will try to explain that dedication, also to myself. One reason (but not the most important one) was that I was given the opportunity to perform 'good physics', as Fermi characterized the atom-bomb programme. Many people thought his statement cynical, but cynicism should not be taken seriously, whereas I believe Fermi meant what he said. We should not forget that the full text of Fermi – "in any case it is good physics" – implies that there is also another side to this work. The physics of atomic and thermonuclear explosions was indeed a "theoretician's paradise".'(*Mijn Leven*, Andrej Sacharov)

'Of course I knew what terrible and un-human things we were busy with. But the recent war had also been a exercise in barbarism; although in that conflict I had not fought, I saw myself as soldier in this new, scientific war (Kurchatov once said that we were soldiers, and to some extent this was not without ground). Over time we have adopted some principles or adapted from others, such as balance of power and mutual destruction, which I still think are in some sense an intellectual justification of the development of nuclear weapons and our role in them.'

'After we had arrived at the test site we learned that an unexpected and rather complicated problem had appeared. The test would take place just above the ground. The device would be exploded on top of a tower built in the middle of the test site. It was known that the explosion above the ground should lead to deposition of radioactive traces; but nobody had thought about the fact that a very powerful explosion, such as we expected, should lead to traces being deposited well outside the test site and thus be a threat to health and life of thousands of people who had nothing to do with our job and who did not even know what was about to happen to them.'

In his memoirs Sakharov leaves no doubt what he and his colleagues were up to: this was a political not a military goal, it was an explosion and not a bomb, a blast instead of a weapon, as weapon it was even useless but the blast had to be so loud it should be heard in Washington.

Back to the question why Teller was despised and Sakharov sanctified, whereas both worked wholeheartedly on weapons of mass destruction? Some people think that it is because Teller betrayed his colleague Oppenheimer, when he accused him of communist sympathies. Although this certainly played a role, it does not explain the position of Sakharov. Teller and Sakharov both fought against the totalitarian communist regime. Whereas Teller sold lies for a 'good cause', Sakharov remained honest until the bitter end. The crucial difference between the two has become public only recently, after the death of Teller. In the August 2004 issue of Physics Today Harold Brown (secretary of defence under Carter) and Michael May (past director of Livermore National Lab) write:

'When asked why he supported Star Wars, a programme with such obvious flaws, Teller answered that if the USA did not work on improving the errors Star Wars should never work. For Teller, whose priority was to defend the USA in the Cold War, this was the only possible stance. But for us, and many others, this was just intellectual dishonesty. Scientists are expected to tell the truth as they understand it and not to make it dependent on some other agenda.'

Through his lies Teller gradually became isolated from his scientific peers; Sakharov was isolated too. Because he was so honest, he was deported by the regime, but his colleagues in the West lobbied for the Nobel Prize. The big difference between Teller and Sakharov is illustrated beautifully by Peter Goodchild in his recent biography of Teller. In *Edward Teller, the real Dr*

Strangelove Goodchild describes the peace talks between Reagan and Gorbachev in Reykjavik, which do not progress because of the Star Wars programme that Edward Teller has talked Reagan into. As long as Reagan does not stop his Star Wars, Gorbachev is unwilling to dismantle his intercontinental ballistic missiles. Then Gorbachev is briefed by Sakharov, that every scientist, also in the USA, knows Teller has sold lies to Reagan and that the Star Wars programme is a big hoax and will never work, so Gorbachev does not have to be afraid and should better offer to withdraw all missiles. Gorbachev does this, to Reagan's surprise, and with the known effect: for the first time in history nuclear weapons are dismantled, thanks to Sakharov. After signing the 'Intermediate Nuclear Force Treaty' Gorbachev and Teller meet with the result given under the title of this text. For Gorbachev Teller is a liar, a scientist who fully well knows he is not telling the truth he does not like, because it does not suit him. In that sense Teller is not alone, for 'There are many Dr Tellers.'

Gorbachev was right. There are too many scientists, also in the Netherlands, who know they are not telling the truth about nuclear energy:

'Tellers' say that a little bit of radiation is good for us humans.
'Tellers' say that nuclear energy is the solution for climate change.
'Tellers' say that nuclear power is cheaper than natural gas.
'Tellers' are against wind farms, for they fear them as alternative to nuclear energy.
'Tellers' say we can look after radioactive waste for many thousands of years.
'Tellers' say nuclear energy does not require a police state.
'Tellers' say nuclear power plants are terrorist-proof.
'Tellers' do not say what to do with the plutonium produced at nuclear power plants.
'Tellers' do not say there is not enough nuclear expertise left.
'Tellers' say they are building a fusion reactor for sustainable development.

If I have learned anything from my teacher it is the heavy duty of scientists to tell the truth, under all circumstances. Perhaps that was the deep reason why we were introduced to Teller.

2004

PS

The 1991 IgNobel peace prize was awarded to Edward Teller "for his lifelong efforts to change the meaning of peace as we know it".

The scientific life

The centennial of the American Physical Society in 1999 was marvellous, with thousands of physicists, countless numbers of scientific contributions, special publications, posters, T-shirts (*"flirt harder for I am a physicist"*), books, plays, films, parties, prizes and dinners, but what I shall never forget is the keynote speech by Hans Mark, past Secretary of Defence and President of the University of Texas, who talked about financing science using a slide showing the American science budget along with the defence budget for the past hundred years. Although the two budgets differed by two orders of magnitude, the two curves ran fully parallel over the whole period of one hundred years. Both showed a steadily growing function with sudden rises (step functions) in 1914-1918 (WW I), 1940-1945 (WW II), 1950-1953 (Korean War), 1956-1975 (Vietnam War). The message was immediately clear to everyone. I was shocked, but the conference hall was enthusiastic because on TV in our hotel rooms you could follow the debate in Congress on President Clinton's plan to bombard Kosovo. That decision would almost certainly be another big boost for the science budget.

'There has never been something like a scientific revolution,' wrote Steven Shapin provocatively in 2004 (*The Scientific Revolution*), but now this historian of science had finally discovered a true scientific revolution which he describes in his book: *The Scientific Life: A moral history of a late modern vocation* (The University of Chicago Press, 2008). If Steven Shapin is to be believed the time that the science budget goes parallel with the defence budget are over. Today the science budget would show the same peaks (and troughs) of the index of the NASDAQ, the American market for technology funds, for in these late modern times scientific life has evolved from truth-seeking to entrepreneurship.

Steven Shapin is Franklin L. Ford Professor of the History of Science at Harvard University since 2004, from 1989 until 2003 he was Professor of Sociology, later also History, at the University of California, San Diego and from 1973 to 1989 at the University of Edinburgh. His latest book, *The Scientific Life*, is primarily based on his experiences in San Diego. He writes modern history and makes the point that he does this in order to better understand 'the way we live now'. With that his work is more in line with sociology of modern science, indeed all chapters start with a beautiful citation from Max Weber's *Science as a Vocation* from 1918. Shapin makes it very clear that he appreciates the dominant role of modern science and technology in our society and he focusses on the motivation and the morality of the scientists in what he calls 'the world of making the world to come'.

The Scientific Life starts by briefly describing the natural sciences in the Western world in those good old days when science was still a vocation, but also a time in which scientists were already recognized as 'common' people. Then the emergence of big scientific research laboratories in industry in America, to which this book is limited, is described. We get to know the specific characteristics, as far as they exist, of scientists in 'Big Science' and the problems that managers of those research labs face with guiding those 'free spirits' while they also have to report properly on the research results to the accountants of the businesses and the government. Finally, via the microelectronics and the biotech revolution we arrive in our modern times of financing via 'venture capital' and scientists as entrepreneurs. It is this revolution in the life and work of researchers that interests Steven Shapin most, a revolution which he describes more with admiration than with a critical eye.

Now I am not a sociologist or historian but I happen to be a physicist who has been actively involved for more than forty years, first as researcher and later also as manager, in the scientific revolution that Steven Shapin describes in his book. I have lived that scientific life. It is from this perspective that I have some comments to make on his 'living history'.

'Big Science is: big funding, big instruments, big industry and especially big government as its patron, and lastly big organizational forms in which science was conducted.' This is from Alvin Weinberg, director of Oak Ridge National Lab and member of President Eisenhower's Science Advisory Committee. Big Science is primarily associated with the Manhattan Project, but the phenomenon Big Science dates, as Shapin makes very clear, from well before WW II even from before WW I. The delivery rooms of industrial production of knowledge stood in the big American power companies in the beginning of the twentieth century: General Electric, Westinghouse etc. In the first decades followed by AT&T, Eastman Kodak, DuPont, Dow, Standard Oil and Minnesota Mining and Manufacturing (3M). It is estimated that the American industry in 1920 spent already $20 million in three hundred research centres. Five years later the budget of Bell Labs alone was $12 million and just before the great economic crisis of 1929 the total expenditures by American industry in research had risen to approximately $130 million in 1000 laboratories. Similar developments took place here, such as the Research Lab of our own Philips, but Shapin leaves Europe out of his study and that is a pity for he may have missed something.

Big Science was not born in the Manhattan Project, but not in the American industry either. Perhaps the cradle of Big Science stood in our country when Heike Kamerlingh Onnes on 19 September 1882 took over the leadership of the physics laboratory at Leiden University. After thoroughly studying the life and work of Heike Kamerlingh Onnes, Dirk van Delft writes: 'Heike's project was Big Science.' (*Freezing Physics, Heike Kamerlingh Onnes and the Quest for Cold*, 2008) 'Anyone who entered the Steenschuur premises, especially lab E and the surrounding area, and beheld the profusion of tubes, taps, gas flasks, gas holders, liquefiers, Dewar flasks, cryostats, clattering pumps and droning engines, glass-blowing and other workshops, instruments and appliances for scientific research, would have felt as if he had come to a factory. It was indeed a "cold factory", with Professor Kamerlingh Onnes as its director, determining policy and exercising tight overall control. As the director of an enterprise, he also set up a well-oiled organization presided over by an administrative supervisor, a research team including assistants and postgraduate students, a manager, instrument-makers, glass-blowers, laboratory assistants, technicians, an engineer, an assistant supervisor, not to mention a small army of trainee instrument-makers to perform any number of odd jobs.' If this is a proper description of Big Science, I do not doubt it and also Shapin should have to admit that Dirk van Delft is right in his conclusion: 'With his Big Science approach, his carefully orchestrated research programme, in which, instead of working individually, everyone contributed to a team effort in pursuit of a well-defined goal, Onnes set an example that other laboratories later emulated.' Such as Philips Research in Eindhoven, which was founded in 1914 following the example of the lab in Leiden with Gilles Holst, pupil of Kamerlingh Onnes, as its first director.

Implicitly Dirk van Delft also corrects the misunderstanding that Big Science would be impossible in a university environment, it does not even belong there, as Shapin writes. He refers to the Nobel committee who awards the prize to a maximum of three researchers. But Kamerlingh Onnes earned the Nobel Prize just because of his Big Science approach at Leiden University and perhaps he was the first, but definitely not the only one. After WW II most Nobel Prize winners in nuclear and high-energy physics were leaders of big international university research groups and facilities. The large nuclear research centres, like Oak Ridge in America, Chalk River in Canada, Harwell in England and FOM and RCN in our country, also were beautiful facilities with large numbers of researchers. After the first nuclear reactors came on the market and the atomic and nuclear scientists had finished their jobs, or thought so (despite fundamental problems with nuclear waste and safety remaining unsolved), the research and development labs were all converted into Big Science centres for fundamental research. It is beyond doubt that the success of the FOM Institute in Amsterdam, first with uranium enrichment and today with fundamental research, is also due to the fact that its founder was educated at the Kamerlingh Onnes Lab. The fantastic research facilities and the extraordinary budgets have put university physics into a privileged position, which the physicists have been able to keep until this very day and not only in this country.

I would like to add, something that Steven Shapin or other sociologists and historians can hardly imagine, that once you have been a scientist in a Big Science environment you never want anything else. Scientific research is extremely costly, especially on the part of the scientist and therefore it is too heavy a load for most people. First you have to convince others of your research plans so that you will get the necessary funds and facilities. Then your research turns out different from what you had anticipated, frequently the first results are useless because you have overlooked something or because the apparatus does not function according to plan. After you have finally discovered something comes the battle to get the results published. For 'normal' people these frustrations are too much. It is then a consolation to work with FOM or any other big research organization where you are part of a group of researchers with whom you may share your sorrow and where there is always someone with a good idea. I do not understand how the lonely knowledge worker sustains in his own little room, deprived from the daily coffee room with his national and international colleagues, without the stimulating discussions and the Eureka effect which in a large laboratory at least once a week causes excitement, without the role model of the research manager.

When writing about the management of Big Science Steven Shapin prefers to cite Kenneth Mees, the first director of Eastman Kodak Research: 'When I am asked how to plan, my answer is "don't". The best person to decide what research work shall be done is the man who is doing the research. The next best is the head of the department. After that you leave the field of best persons and meet increasingly worse groups. The first of these is the research director, who is probably wrong more than half the time. Then comes a committee, which is wrong most of the time. Finally there is the committee of company vice-presidents, which is wrong all the time.' How funny this may sound, I wonder what Kamerlingh Onnes would have said, for Mees ignores the mission and orchestrated approach, the two most important criteria of Big Science. On Kamerlingh Onnes as manager Dirk van Delft writes: 'This "Big Science" approach, unique in its combination of focus and the large scale on which everything was tackled, could only succeed with someone at the helm who had persistence, courage, willpower, vision, and indestructible patience. Someone who ruled with a firm hand, but who at the same time had a gift for winning people over, persuading them, securing their loyalty. And someone with peerless ability to manipulate the powers-that-be – Heike was always warning that what had taken years to achieve at the Steenschuur was in danger of being destroyed – and he kept up his dire warnings until the authorities gave their "expensive professor" the space and the resources he needed to accomplish his goals. He was a brilliant networker, with a keen eye for useful contacts both within and far beyond the field of physics, someone who pampered his guests and was far too shrewd to quarrel or to make enemies who might harm his interests... Onnes was a sound scientist, but his cryogenic laboratory owed its success to his talents for organization, his social skills, and his unswerving focus on extremely low temperatures.'

The norms and values of the scientist form the red thread that runs through *The Scientific Life*. In the beginning of the twentieth century the scientist was still a lonely explorer in search of truth, perhaps exceptionally talented and motivated, but a normal human being as far as morality is concerned. Upon the new developments in the first half of the twentieth century and with the transition to the industrial production of knowledge, a new tension occurred between the motivation of the scientist and his sponsor, both in industry and with the government. What would be left of the moral responsibility of the researcher, who became only a small speck within a Big-Science organization? For good reasons WW I may be called the war of the chemists, whereas WW II may be called the war of the physicists. Some older physicists from the Manhattan Project had returned to their universities with a bad conscience, but the younger generation did not worry to show up at the gates of Oak Ridge, Los Alamos and Livermore, to contribute to the Cold War weapon development under the guidance of Alvin Weinberg, Edward Teller and Hans Mark. To them Eisenhower's farewell speech, in which he warned us for the dangers of the military-industrial complex, fell on deaf ears. How very different things were in our country during the days of the Cold War.

People like Casimir, and his co-worker Dippel of Philips Research, were up front with discussions on the social responsibility of the scientist.

Steven Shapin only reports about developments in the US, without much judgment. He states that the number of Nobel Prizes which went to Bell Labs after WW II indicates that there was ample opportunity for fundamental research in industry. On the other hand, he hears from Mees (Kodak) that most scientists in industrial labs were so much impressed by the mission of the industry they worked for that it determined all their norms and values. Mees frequently had to remind them of their own creativity. The fact that at Philips under Casimir moral responsibility was considered important also indicates that not everyone was made to contribute to the same goal. At Philips Research there was an active department of the Verbond van Wetenschappelijke Onderzoekers (under Dippel) and later in the 1960s the Bond van Wetenschappelijke Arbeiders, a leftist separate movement of concerned scientists. The 'laissez-faire' mentality of managers such as Mees and Casimir and also the leaders of Bell Labs, have created the second generation of Big Science, still with a big research organization and fantastic facilities but without a clear mission.

The absence of one clear goal, the 'degenerated' form of Big Sience, has weakened those laboratories and is probably the cause for one of the most important developments in industrial research in last quarter of the twentieth century, which Shapin completely ignores. The retirement of the generation of research managers of the calibre of Casimir and Mees has meant the definitive end of fundamental research in industry. The financing of the big research labs would no longer come from the Board of Directors, but would be dependable on orders from the industrial divisions. The scientists were expected to go 'begging' for money with the production units of the company. This fundamental change in the management of research in industry looked like it was orchestrated as it was enforced almost simultaneously all over the world: at Kodak, Bell Labs, GE, Dow, and also at our Philips, AKZO, Unilever and Shell. The effect was to be expected: an accelerated introduction of new technology to the market, but also the departure of some of the best scientists, no further Nobel Prizes and even the closure of Bell Labs in the US and of Shell Research in our country.

In his book Shapin frequently wonders how 'for heaven's sake' science is planned. The results of fundamental research are by definition unpredictable, let alone the time frame within which the results should be reached and the budget needed. This seems reasonable, but we researchers have readily learned that trick of writing research plans. In fact, it is not as difficult as it seems: your most recent research results, the ones you have booked but not yet published, you submit as a research proposal for the coming year. Discoveries always generate new questions and you submit those questions the year after that – it is very simple – and by the time your latest proposal is up for review the first results of your previous investigations go to press. Success assured.

We have learned one more 'trick'. Scientists and engineers in the micro-electronics industry, from the large technological institutes and from the universities, made sure that the computer chip would become a hype years ago. We went to our bosses in industry, the national labs and the universities and we argued for investments in new production facilities, new laboratories and new colleges, because the microchip was coming. We went to the media and we explained what that new technology would mean to our society. We went to The Hague and told the politicians about the new economy that was coming and that we would miss the boat. We went to Brussels and asked for stimulus programmes to make sure Europe would stay in competition with the US and Japan. And it has all worked out. Because we worked together they believed us and we received new investments for new production facilities, for new laboratories and for new colleges, for new students. Thus, we from the industry and from the national labs and from the universities have created and built the microchip technology. It worked because we all worked together.

But let us not forget that networking, first in computer technology, then in biotech and now in pharma, is an American innovation. They were unique because they were financed by 'venture capital', something that did not exist in Europe. The VCs, the 'Business Angels' and the NASDAQ have meant a true revolution in what Steven Shapin calls 'techno-scientific knowledge'. Private capital has made possible in America what was impossible elsewhere. It gave a magic ring to Silicon Valley in California and Boston Route 128. Shapin estimates that in America, before the financial crisis, there were as many as 3500 VCs who together invested 100 billion dollars in 'techno-science'. When I see such amounts, I do not understand why on the balance sheets of scientific organizations you see listed their money, buildings and debts, but not the value of their knowledge. According to Shapin only one out of ten investments does not deliver anything, three keep the same value as the amount deposited but not more ('the walking-or-living-dead'), two have a revenue of 200-300 per cent and two become worth more than ten times the original investment. In the last category are of course the chip manufacturers Intel and Applied Materials, the software giants Microsoft and Google, the biotech companies Genentech and Celera. Most dot-com companies went broke when in 2000 the internet bubble burst.

That year was the record year so far with 104 billion dollars in 7813 'VC deals', but we all have heard other success stories since then: MySpace, Facebook and YouTube. A famous VC from California told Shapin about Mark Zuckerberg, the founder of Facebook. He offered this entrepreneur a drink in a bar, but Zuckerberg was not yet twenty-one: 'I had a glass of Pinot Noir and he had a Sprite.'

How do VC deals get done? In 2007 as much was invested in national security and military technology as in green energy. What drives a VC? It is earning money, a lot of money and nothing else. But what drives the scientist to become an entrepreneur? And how does the one find the other? For Shapin these are the most important questions in his *The Scientific Life*. He spends two chapters plus an epilogue on them. First, he poses the question from the point of view of the VC: 'What do VCs think they're doing when they think they are doing pretty well? And how, so far as one can judge, do they actually confront the radical uncertainties of the world in which they've chosen to make their living?' Shapin philosophizes about this for a full chapter, but finally it boils down to the fact that the VC only strikes a deal if he is convinced of the commitment of the entrepreneur. Thus the question remains: What drives the scientist/entrepreneur? Another full chapter Shapin philosophizes about this, whereas you would think the answer is simple and that money plays the major role. Finally, in the epilogue about networking on a patio under the palm trees of California with an ocean view and a glass at hand, Shapin pulls the rabbit out of the hat: 'It's not all that rare to hear people spontaneously say that they're trying to "make the world a better place" and that they're committed to wiping out some dread disease.' And also: 'They're in the business of techno-scientific and economic future making, trying to discover drugs that will cure or alleviate cancer, or wireless technologies that may become world standard...'

When I read that, I thought: Should we try again? Shall we go to the companies again, to the government and to the universities to start a new hype? What shall we choose this time? Shall we now go for sustainable development? Shall we go to our bosses again and ask for investments for sustainable technology? Shall we go to the media and explain what sustainability means to society, and plead with the politicians in The Hague and Brussels for sustainability in the interest of the climate and the economy? Yes we can.

2009

To colleagues and friends, for decades of Fun,
Utilization, **Theories of everything** *and Survival,*
Thank you

Shopping

'Applied science is like shopping with a shopping list made at home before. One has already decided what is needed, but nature hides secrets of which we don't know they exist. To find them we have to look around quite extensively without a shopping list.' That is how Gerard Nienhuis defended the primacy of fundamental research in his inaugural speech on 'The unity of physics' at Utrecht University.

In this speech Gerard was brave enough to pose the question: 'Can we say that progress in physics has brought us closer to reality?' A venture for the answer might be embarrassing to us physicists. Gerard illustrated progress in physics using six examples of very diverse phenomena, which at first glance do not seem to be connected but which at a deeper level may all be recognized as different manifestations of the same effect. Could it be that a deepest level might be found on which the entire scale of physics phenomena can be put?

It was Einstein's conviction that it was the physicist's task to uncover this deepest level. But physics research of the twentieth century is characterized, to the great disappointment of Einstein and his disciples, by the impossibility to seize reality. 'The more the physicist's description of nature goes down into deeper levels, creating the expectation that one gets closer to reality, it turns out that reality gets further out of sight. It seems as if reality shifts back when we get closer, like the horizon is moving back for the traveller, just as the end of the rainbow or the vision of the unattainable loved one.' In a very lucid way Nienhuis described the strange situation in which the modern physicist finds himself. Another text in which the crisis in physics is put so well, will not be found easily. This makes it even more amazing that he remains true to his ideal: 'If we give up an ideal, only because it is unattainable, then we will never know how close we might have come.'

I think that is like shopping in a supermarket trying to find the elixir of life (and then feeling close near the liquor section?).

Would it not be wonderful indeed if the physicist's ideal could be attained? If it were true that the next generation of particle accelerators would create the fundamental building blocks of nature and the next generation of computers their fundamental equation of motion? Sadly enough this will not be the case. If progress of modern physics has also created progress in our understanding, then it is this very realization. Realizing that we will neither find elementary particles, nor fundamental basic equations. What then urges us to hold on to the ideal?

Nienhuis has two reasons for it. In the first place there is the fear that physicists without the ideal of an objective reality will get out of control and will follow different motives than merely curiosity. Such as chasing what is in fashion and to start competing for the sake of competition or out of jealousy. These are the people who go shopping to fill up the shopping cart as full as possible in the shortest possible time, preferably in front of TV cameras. It does not take long roaming about in physics to recognize that this is not an unimaginable danger.

In the second place, according to Nienhuis, there is the widespread belief that fundamental research precedes applications. And it sounds reasonable, but it may only seem true, for when you go shopping merely to shop you might also come across something useful or practical.

Why not change paths and collect knowledge in the first place to do something useful with it? Then we are no longer out of control and we do not have to fool ourselves and others; we are going shopping, not to find our elixir of life but to provide in our daily necessities of life. It goes without saying that this ought to be well organized by making a shopping list in advance; that is OK. What is not OK is walking around with blinkers. It may well be that eventually it turns out that we have found more than we had put on our shopping list and also that for the time being some of our wishes have not been fulfilled, which makes more fundamental research necessary. But then at least we know why.

1985

Fundamental research on matter

With an enormous bang our universe was created some fifteen billion years ago. What it was going to look like had already been decided in the first three minutes. After that the entire number of nuclear particles – protons and neutrons – would not change anymore. The total charge of the universe also would be conserved, as well as the number of electrons, muons and neutrinos. Add a huge quantity of photons – particles of light – mix them well, let it expand and after a very long time the original soup will without a doubt change into our universe as we know it, including the Milky Way, the sun, its planets and us.

According to geologists the crust of the earth was formed five billion years ago. The oldest microorganisms found in there are four billion years old. That was when for the first time life on earth came into being from dead matter. Is the entire quantity of life in our universe perhaps also a number that is conserved, like the total number of protons and neutrons, or is new life still being born from dead matter?

With special telescopes the signal from complex organic molecules is detected. Simple molecules of hydrogen or carbon monoxide may be formed if it so happens that two atoms hit each other and as a result stick together. But it is highly improbable that nine different atoms would do that to form ethanol. Yet these and other organic molecules have been detected and people wonder how they have come about.

In between the stars big clouds of dust are observed. In due time, lots of atoms and simple molecules will stick together on some of these dust particles. What may happen if such a particle comes close to a star, we have recently simulated in a lab experiment. The surface of an interstellar dust particle was imitated by a very cold surface in vacuum. On this surface we froze a mixture of simple molecules like water, nitrogen and methane. The presence of a star was simulated, by firing the cold target with a beam of hydrogen ions from a small particle accelerator. It is well known that apart from lots of light our sun also sends a large quantity of hydrogen ions in our direction, but these are caught by our atmosphere. Dust in space does not have that protecting blanket, so consequently it gets steadily irradiated. In the laboratory we could measure with a mass spectrometer (sorting machine for molecules) which molecules are sputtered from the surface due to the hydrogen bombardment.

The first results with frozen water molecules were disappointing for we only found hydrogen and oxygen. However, when we froze methane on our target and bombarded it with hydrogen ions, we suddenly observed the emission of very large organic molecules. Apparently, by irradiation with hydrogen ions it is possible to break the bonds of the methane. The broken bonds of neighbouring molecules connect to form long chains of hydrocarbon molecules. These create so much heat that the newly formed organic molecules come loose from the surface and move about freely into the vacuum. We measured a speed of 100 km/sec. The efficiency is not bad, for every Hydrogen ion that hits the frozen methane a new molecule gets into the vacuum.

No wonder that large organic molecules are found among interstellar dusts. Closer to earth, in our solar system, this process should also take place. The most distant planet, Pluto – and its moon Charon – do not have an atmosphere and they are covered with a layer of frozen methane. So, from here the solar wind should make a lot of organic matter that gets into our solar system.

Does this mean that physicists have discovered the origin of life? Of course not, the tendency to relate fundamental research on simple reactions with elementary particles to the origin of the universe, does lead to a lot of publicity but underestimates the complexity of the evolution from the big bang till today.

1991

A vacuum is not nothing

Even with modern technology it is still not possible to pump all atoms and molecules away. Moreover, according to modern physics, the vacuum between and within atoms is not really empty. What then is a vacuum?

Probably it was Torricelli who made vacuum visible for the first time. He filled a glass tube of a meter long and closed it on one side with mercury and put it carefully upside down into a bucket. The mercury went down and left an empty space in the top of the tube, the vacuum. The mercury did not leave the tube completely. A column 76 cm high remained inside. This was kept high by the air pressure. We know this effect from washing the dishes. If you pull a glass out of the water upside down, no matter how large or high the glass is, the air pressure always keeps the water inside the glass. Usually we do not appreciate how high the air pressure is. That is why Otto von Guericke did his famous experiment with the Maagdenburger hemispheres, which sixteen horses could not separate after the air had been pumped out. The air pressure on a square meter represents a weight of 10,000 kilograms. We do not notice until there is a vacuum somewhere, for instance in a vacuum-sealed coffee bag. The coffee is pressed together because the bag cannot give any counter-pressure.

In the top of Torricelli's tube, inside the Maagdenburger hemispheres and in the vacuum-sealed coffee bag, the pressure is not zero. There are still very many atoms and molecules present in the vacuum. In the Maagdenburger hemispheres after pumping still 1 per cent of all molecules were left, or 10exp17 molecules per cubic centimetre. Today that is called low vacuum, in contrast to high vacuum, in which the density of molecules still is 10exp10 per cubic centimetre. For industrial applications low-vacuum is generally enough.

One should not only think of the packaging industry, but particularly the high-tech industry. Such as microelectronics, where thin films of very high quality materials are deposited in a vacuum, for the production of computer chips. Or in the metal industry, where steel tools are hardened and coated, inside vacuum furnaces. In chemical technology vacuum is becoming more and more important if one wants to work with ultra-pure materials. The largest vacuum in the Netherlands is in Almelo at Ultra Centrifuge Nederland (UCN), where uranium is enriched in a cascade of gas centrifuges. Although in vacuum, enormous quantities of gas are being processed. At UCN they blow gas inside their vacuum, so it is anything but empty.

High vacuum is used in scientific research. With modern analysis techniques the composition and structure of materials is made visible on atomic scale, but then foreign molecules, such as air and water vapour, should be out of the way. Therefore in electron microscopes, mass spectrometers and other measuring equipment increasingly higher vacuum is used. The most extreme vacuum conditions are required for particle accelerators built for high-energy physics. In Geneva there is a circular vacuum tube underground with a length of twenty-seven kilometres. The particles that rush around should not collide with any atoms on their way in the vacuum. Therefore the tunnel is permanently kept in an ultra high vacuum (i.e. ultra low pressure).

The best pumps are in reality freezers; they operate at low temperatures and freeze most atoms and molecules from the air, but not all. A cryopump at a temperature of -200 °C is excellent for pumping water, at -250 °C all gases are frozen except neon, helium and hydrogen. To reduce the helium vapour pressure below 1 atmosphere the temperature has to be reduced still further, close to absolute zero. Therefore in ultra high vacuum systems one always uses a variety of vacuum pumps. For the pumping of noble gases mechanical pumps or ionization pumps are used next to cryopumps.

The best vacuum that could be achieved on earth in this way is still not empty. One has to count on the presence of at least a thousand atoms and molecules per cubic metre. There must be a lot of room between the atoms. Is that space empty? Outside our atmosphere, there are undoubtedly areas where one can find at most one hydrogen atom per cubic metre. Is the space between those hydrogen atoms real vacuum; is that vacuum truly empty?

According to Einstein matter is merely a form of energy. If we would empty a volume by pumping away all matter, but we would leave energy behind, in whatever form, then this can appear again somewhat later as matter. In other words: where there is energy the vacuum is not empty.

In modern physics one goes one step further and assumes that the vacuum is filled with an infinite number of electrons and positrons (the antiparticles of electrons, with the same mass but positive charge). This assumption was necessary in order to unify the relativity theory with quantum mechanics. Although intuitively an infinite number of particles seem contradictory to vacuum, recent experiments are in accordance with it.

A few examples: an atom in a sea of electrons and positrons will behave differently from an atom in empty space. Very close to the positively charged nucleus the negatively charged electrons of the vacuum are attracted and the positively charged positrons are repelled. This polarization of the vacuum must have an effect on the electrons that belong to the atom, which has indeed been observed. Very close to the nucleus the atomic electrons experience the vacuum polarization and get just a little extra energy than in free space. With modern spectroscopic instruments this energy difference has been detected.

One may also wonder whether electrons and positrons in vacuum may be split completely and detected separately? In experiments in which two heavy atomic nuclei are collided against each other such that the nuclear barriers touch, the electric field strength is locally high enough to liberate positrons. In these experiments positrons have indeed been observed.

In empty space two rays of light can pass without hindering each other, but in a sea of electrons and positrons a beam of light will, if it is strong enough, influence the distribution of negative and positive charge in space. Therefore two intense laser beams will deflect each other because of their vacuum polarization. This effect has been predicted long ago but could not be detected due to lack of light intensity. In different labs in the world, one soon hopes to have enough laser intensity to see light collide with vacuum.

Empty space I can imagine, I only have to think of Torricelli's tube and the hemispheres with the sixteen horses, but a vacuum filled with an infinite number of electrons and positrons, honestly I do not understand that at all. Perhaps it is only a matter of getting used to it, for seeing is believing. That was also true for the citizens of Maagdenburg.

1991

Father of the atom

Most important theories in physics I do not understand. At university we studied the Maxwell theory and we learned to calculate the magnetic field that goes with an electric current, but why magnetic north and south poles attract each other, or what a magnetic monopole is, I still do not understand. I have studied relativity theory and passed the test, but why two atomic clocks differ if one is sent with a rocket, I still do not understand. I cannot even remember which of the two clocks is slow. It is the one in the rocket, I believe, but to be sure I would have to make a calculatation. This is somewhat strange for a physicist and that is why I buy books like *Subtle is the Lord*, the biography of Einstein, written by our fellow countryman Abraham Pais.

The opening chapters of this book are most fascinating; despite pedantic talk about walks Pais took with Einstein. The state of physics at the end of the nineteenth century and beginning of the twentieth century is described brilliantly. For me Einstein's work and $E = mc^2$ had come out of the blue. Now it turned out that Einstein struggled, like most of his colleagues, with the ether theory. While worrying about that the relativity theory appeared. Ether disappeared into the background, at least for a while. I did not know Einstein's earlier work, which was mainly on statistical physics, and his research on Avogadro's number is extremely interesting indeed.

What comes after that, the real subject of Pais's book, has remained unintelligible to me. Too many formulas, too much implicit statements and conclusions. Pais is a typical theoretician who can think and associate mathematically, someone for whom the beauty of mathematics becomes clear in the wink of an eye. I read through a bunch of formulas, sometimes skipping pages with 'nothing' in them. Still I read on, for the passages on the personalities of Einstein and his colleagues were exciting. And I remained hopeful about learning more about the essence of the relativity theory. It did not work.

It is a consolation that I am not alone, as is apparent from the book of Pais. The London Times of 7 November 1919 writes about a meeting of the Royal Society in connection with the experimental verification of Einstein's prediction: 'No speaker succeeded in giving a clear non-mathematical statement of the theoretical question.' Extensive coverage is given to the enormous interest in Einstein's work, not only from his colleagues but also from the general public, which does not understand anything but still crowds around to get hold of his latest 'paper'.

I feel angry and I blame Pais for not really explaining relativity to me. Then suddenly there is the statement from Hertz: 'Maxwell's theory is Maxwell's system of equations.' Pais calls this a funny but useless comment on the best of physics at the time. For me this statement is of unusual importance because more and more I get the feeling that 'big' physics is no more than a set of mathematical equations with which experimental results are predicted, but of which the deeper meaning escapes us. This is not only true for the equations of Maxwell, but also for the relativity theory of Einstein, as well as for the quantum mechanics of Niels Bohr.

To see if Abraham Pais agrees with my view, I have eagerly started reading his most recent biography: *Niels Bohr's Times in Physics, Philosophy and Polity* (also Oxford University Press). Niels Bohr's philosophy is summarized by Pais in the first chapter. 'Quantum mechanics renders meaningless the question: Does light or matter consist of particles or waves? Rather, one should ask: Does light or matter behave like particles or waves? That question has an unambiguous answer if and only if one specifies the experimental arrangement by means of which one makes observations.' Or better still in Bohr's own words: 'Our task is not to penetrate into the essence of things, the meaning of which we don't know anyway, but rather to develop concepts which allow us to talk in a productive way about phenomena in nature.' Then follows a text that is at least as fascinating as the biography of Einstein, perhaps even more so, because no one else but Niels Bohr combined three properties in one. He was a 'creator of science, teacher of science, and spokesman not only for science per se but also for science as a potential source for the common good.'

The image we have of the atom, with a nucleus and electrons circling around, is due to Niels Bohr for which he earned the Nobel Prize. Yet this image is weird, because if electrons circle around in fixed orbits in a circular accelerator they radiate light and will circle more and more slowly as they lose energy, thus we have to keep accelerating the electrons to keep them in orbit. Atoms are not small accelerators; still the electrons inside occupy stable orbits. This is explained only by assuming that the electrons inside the atom behave like waves, but that is strange, is it not? Indeed, that is 'the strangeness of quantum mechanics'.

This weird discovery by Bohr had an enormous attraction for his fellow scientists. The institute in Copenhagen that was named after him, and the castle of the Carlsberg Foundation that was given him to live in, became the centre of modern physics. Between 1916 and 1961 no less than four hundred forty-four scientists from thirty-five countries stayed there. Together they published twelve hundred scientific papers, among which two hundred from or with Bohr. He dominated the atmosphere in his surroundings. In his digging to the very bottom of things he was unsurpassed. He was energetic enough to hold on till the very end. In addition, he was able to enjoy life in general. This made him the father of atomic physics and he treated his pupils as if they were his children.

But like a father he could also make his colleagues desperate, such as Slater ('I had a horrible time in Copenhagen') or Heisenberg ('in tears because I could not stand this pressure from Bohr') and Mott ('I wished Bohr let me get on with it without examining everything'). For most visitors, however, Copenhagen was a turning point in their life, a legendary time, also for Abraham Pais who with great joy remembers the many anecdotes that he describes so eloquently. Sometimes posturing, with words like: polity, unbeknownst, bequeathed, largess, ventripotence, cognoscenti, opine, but always balanced by a complete absence of formulae, much in style with Niels Bohr himself ('we are suspended in language').

Niels Bohr became more and more famous. He was a celebrity among heads of state and premiers, first at home but after WW II all over the world. Thus he was able to acquire funds in Denmark for physics and to found several institutes. During WW II he mobilized his whole institute and produced six thousand gasmasks in a week, just in case Denmark was attacked using poison gas. Later Bohr (half Jewish) had to flee, via Sweden to England and on to the US, where he became advisor to the nuclear-weapons project. After WW II Niels Bohr pleaded for complete open access to all knowledge on nuclear energy and bombs in the interest of avoiding the Cold War. But the politicians thought him naïve or did not understand. Churchill 'would always be honoured to receive a letter from Professor Bohr but hoped it would not be about politics.' Eisenhower on 24 October 1957 awarded him the first Atom for Peace Award, but a week later on 1 November the first hydrogen bomb exploded in the Pacific.

As a politician and strategist in science Niels Bohr was most successful. Yet this complete success was 'bittersweet' or as Hendrik Kramers, Bohr's closest collaborator at the time, once said: 'The quantum theory has been very much like other victories; you smile for months and then you weep for years.' Bohr brought about a truly remarkable synthesis between atomic spectroscopy and chemistry, but the way in which this was accomplished and its consequences meant a true revolution in the natural sciences. The principle of action and reaction was lost. This price had to be paid in order for quantum mechanics to help us where classical mechanics had failed.

Quantum mechanics tells us of radioactive atoms, that today one atom will decay, tomorrow another and the day after one more, but which atom it will be no calculation can predict because quantum mechanics only calculates a probability. Abraham Pais describes in detail the role of all players at the conception of this new gambling game: Planck, Pauli, Einstein, Sommerfeld, Schrödinger, Heisenberg, Born and De Broglie. They all contributed, but it was Bohr primarily who was the architect/contractor who kept his eye on the foundations. Therefore today we speak of the Copenhagen Interpretation of quantum mechanics.

This is a pragmatic interpretation: the only reason quantum theory holds, is because it makes the proper predictions about experiments. This does not mean there is a quantum world out there. There is only an abstract mathematical description. Quantum theory, that is the wave equation. Why? Because it works.

Many physicists, including Pais, are unhappy about that. Bohr accepted it, probably because his predecessor in the castle of the Carlsberg Foundation was the pragmatic philosopher Höffding. In addition, Bohr must have read Steno, the Danish scientist from the seventeenth century, who wrote: 'Beautiful are the things we see/ More beautiful those we understand/ Much the most beautiful those we do not comprehend.'

1991

Who influenced Bohr?

Even today there is still controversy about how much Niels Bohr was influenced by philosophers, if at all. The idea that he was influenced is dismissed as outrageous in the physics world at large. The day before he died, however, Bohr admitted to Thomas Kuhn, the historian and sociologist of science, that he had read the philosopher William James, even though he could not remember when.

Kuhn was interested in the debate between those who view the development as solely depending on the discoveries and discoverers within physics, and those for whom this view is far too limited. The development of science, say the latter, depends not only on the discoveries in a particular field of research, but also on sociological and cultural circumstances outside that field.

Bohr's greatest contribution to science was the Copenhagen interpretation of quantum mechanics. The physics community was in disarray when quantum theory replaced classical mechanics at the beginning of the twentieth century until Bohr showed the way forward. Several authors have noticed that Bohr's is a pragmatic interpretation that should be attributed to influences on him from philosophers like Kierkegaard, James and Höffding – a view opposed by many physicists. In many quarters there is a feeling that physics does not need philosophy, and that Bohr would somehow fall from his pedestal if he had read, and was influenced by, philosophy.

Before the Copenhagen interpretation physicists were bewildered by the uncertainty principle and wave-particle duality. Max Planck, for example, was a religious man who could not accept that the uncertainty principle could also be valid for the Almighty. To Planck the electron was a particle and the wave-like behaviour observed in the two-slit experiment could be explained by God being able to see which slit the electron went through on its way to the detector, without disturbing its path. Of course this is metaphysics, and in conflict with experimental observations.

Bohr's book *Atomic Theory and the Description of Nature* (Cambridge University Press, 1934) opens with the sentence: 'The task of science is both to extend our experience and reduce it to order.' Although he writes such complicated sentences that it becomes almost impossible to read, Bohr's interpretation of quantum mechanics is simple. He says that we cannot answer the question: 'Is the electron a wave or a particle?' And, unlike Planck's God, we cannot tell which slit the electron passed through. The only thing one may expect from the physicist is a clear description of what he has measured. In one experiment the electron will behave like a wave, in another it will behave like a particle, but the electron is neither. It is an 'electron', and this word stands for all our experiences in experiments with electrons.

Later Bohr wrote (in *Atomic Theory and the Description of Nature*): 'In our description of nature the purpose is not to disclose the real essence of phenomena but only to track down as far as possible relations between the multifold of our experience.' In Holland we simply say: 'To measure is to know.' According to Bohr we do not know through which slit the electron goes because that is not what we set out to measure. The same is true for the light in the refrigerator – we think it is off when we close the door, but we will only know for sure when we climb into the fridge and close the door behind us.

In classical mechanics we know the position and velocity of a number of particles and we can calculate the trajectory they will follow. In quantum mechanics we can also predict the future, but in quite a different way. We know the initial condition of our experiment and we can calculate the most probable outcome. Although the result of the experiment is accurately predicted by quantum mechanics, it does not tell us how the particles move from the initial to the final state. But this is not necessary because this is not what we measure. As Bohr wrote: 'Strictly speaking, the mathematical formalism of quantum mechanics and electrodynamics merely offered rules of calculation for the deduction of expectations about observations obtained under well-defined experimental conditions specified by classical physical concepts.'

But should Bohr have referred to William James? In philosophy pragmatism is defined as a doctrine that evaluates any assertion solely by its practical consequences and by its bearing on human interests. Nobody denies that Bohr's interpretation of quantum mechanics is pragmatic, and that he was the first to say that problems of modern physics could be dealt with in this way. But Bohr did not discover pragmatism. For this we must turn to the philosophers, especially William James.

In *Pragmatism* James wrote: 'Theories thus become instruments, not answers to enigmas, in which we rest... They are only a man-made language, a conceptual shorthand, as someone calls them, in which we write our reports of nature; and languages, as is well known, tolerate much choice of expression and many dialects.' This was in 1907, long before Bohr pointed his colleagues in the same direction.

The interview with Kuhn makes it clear that Bohr was familiar with James's philosophy early on – even before he went to Manchester as a post-doc in 1912. He should thus have referred to James in his writing. Rosenfeld, one of Bohr's colleagues, tried to persuade him to mention the affinity between his approach and that of James explicitly, but Bohr refused to do so, 'not because he disagrees, but because he intensely dislikes the idea of having a label stuck onto him.' (Stapp 1972: 1098)

It would have been better if he had referred to James. It would have done justice to the philosopher. Moreover, Bohr need not have been ashamed of his knowledge of philosophy and other disciplines. On the contrary, Bohr was able to show his fellow scientists, including Einstein, the way in modern physics because he was not only interested in physics but also familiar with other disciplines – contemporary philosophy in particular.

1995

Hooray for the electron

We have started to celebrate the centennial of the electron. According to some, the birth of the electron coincided with the Big Bang, fifteen billion years ago, but we physicists got to know the electron through J.J. Thompson's experiments with cathode-ray tubes. In 1895, he demonstrated that in what we today would call a TV tube, a beam of particles travels from the negative side (the cathode) to the positive side. Thompson determined their charge and mass and baptized the carriers of electric current: 'electrons'. These experiments did not prove that all electrons are identical and some anti-atomist physicists argued that the electron is not a unique particle and that the measured current represented only a mean value. Therefore Millikan did experiments with droplets of oil to which he attached charge and thus made them to float in an electric field. In this way he could determine that the droplets were always charged with a whole number (two, three, four) times the charge of Thompson's electron. By adding a grid in the electron tube Lee De Forest could control the electron current and even amplify it. This led to the birth of modern electronics and the development of radio, TV, radar, computers, electron microscopes, etc.

When everybody was convinced that the electron was a particle, a fundamental building block of nature, the atomic physicists spoiled the game. To explain the behaviour of the electron inside the atom, one had to assume that it was not a particle but a wave. Proof of the wave nature of the electron came from Davisson and Germer, who scattered electrons of a nickel crystal and observed the same interference patterns as x-ray diffraction produced. Yet, it was strange that in the one experiment the electron behaved like a particle and in the other like a wave. It became stranger still, for the electron turned out to rotate around its own axis, which is why in a magnetic field it may stand on its head or tail. It was also predicted and verified that the electron has an antiparticle, the positron, with the same mass as the electron but with opposite charge. If an electron and positron collide both disappear and what remains is merely a flash of light.

The image we have today of the atom consists of a positive nucleus surrounded by a cloud of negatively charged electrons. The nucleus determines the mass of the material and the positive charge attracts the negative electrons to a particular position in the material. Almost all other properties of matter are determined by the electrons, not by the nucleus. Whether a material is hard or soft, gaseous, liquid or solid, blue, yellow or red, an insulator or a conductor, the structure of the materials around us is all determined by the electrons. It is the atomic physicists dream to know the behaviour of the electron so well that, based on this, all possible macroscopic properties of matter may be explained. Atomic physicists are really electron physicists rather than nuclear scientists.

In 1984, physicists at AT&T Bell Labs learned to control the behaviour of the electron in a crystal just a well as in Lee De Forest's electron tube. They built a miniature electron tube, the transistor, from which microelectronics would emerge; a completely new industry. Today microelectronics dominates modern technology. No other industry is so innovative and spends so much money on research and development as microelectronics. No wonder, for the transistor, computer, radio, TV, telephone, radar, CD, microwave, superconductor, solar cell, video, Walkman, mobile phone, fax, email, and all sorts of medical and scientific instruments and technology, you cannot pump out of the ground like oil. They have all been developed in the research labs of the electronics industry, which soon should become equal to or larger than the oil or chemical industry.

In preparation of the centennial of the electron Philips has shown its latest discovery: the high-definition television. There were also some surprises, however, from the scientific community. In a collaboration between Philips and the Technical University Delft it has been discovered that transport of electrons through a narrow gate is quantized. If the gate through which the electrons go is narrow enough, equal to the size of the wavelength of the electron, interference will occur. If one opens the gate just a little bit more, then there will not immediately be more electrons passing through. Only if the width is a whole number of wavelengths of the electron will the current jump up. This discovery may lead to a whole new class of electronic devices based on the wave character of electrons. The same group in Delft has also developed some kind of revolving door for individual electrons. In a special electronic circuit the electrons are, as if they were particles, transferred one by one from one place to another in the circuit. The FOM Institute in Nieuwegein, the Netherlands, demonstrated its 'Free Electron Laser', an accelerator in which fast electrons are sent through a strongly varying magnetic field, thus tuneable laser light is generated in the far infrared part of the spectrum. The National Institute for Nuclear and High Energy Physics in Amsterdam demonstrated a 900 MeV electron accelerator, a kind of electron microscope for atomic nuclei, with which the behaviour of quarks in different nuclei may be compared. At the FOM Institute for Atomic and Molecular Physics one can see electrons circle around the nucleus.

The anniversary of the electron was badly disturbed by new measurements of the American Nobel Prize winner Dehmelt. Whereas everybody 'knows' the electron is infinitely small, Dehmelt started looking for its radius. He built a trap in which he could put the particle and observe it for quite some time. Preliminary results showed that the magnetic properties of the electron differed from the theory of infinitely small particles. From this he concluded that the electron has a radius of a magnitude of 10exp-21 cm. Whatever the size of the electron, what is it made of? How is it possible that it may behave both as a particle and as a wave? How can an electron have mass but no size? Is the anti-particle, the positron, just as large or smaller? How can they annihilate one another in a flash of light, whereas both are material objects?

At its centennial the electron had a special present for us physicists. After one hundred years of fundamental research on the electron we should know what the proper question is we physicist should try to answer. Not: What is it? Or: What is it made of? After a hundred years we still do not know and probably we will never know. Instead we should ask: How does it work? That is the fundamental question. Answering that question has changed our society fundamentally.

1992

Decadence

We live in a decadent world. And we know it. In politics, in sports, on television, in the arts and in the sciences, decadence prevails everywhere as if it knows the twentieth century is already at its end (as we write it is 1999). The end of the nineteenth century, the 'fin de siècle' is known for its decadence except in my field, physics, but today it is unmistakeable, also in physics. It does not have anything to do with the new millennium, but let me use this special occasion to ask the question from where the prevailing decadence in physics has come, whereas it was absent during the turn of the last century.

Physics has fallen apart into a large number of specialities like: elementary particle physics, nuclear physics, atomic and molecular physics, statistical physics, solid state physics, bio- and medical physics. Most physicists spend their entire life on one such discipline. It is almost impossible to change fields because it usually takes many years to learn enough of a particular specialism to be able to make a creative contribution. In addition, most fields of research require enormous investments to stay in line with the international competition. Once the investments have been made it is imperative to be as productive as possible and thus the publication machine once started cannot be stopped. A production of ten scientific publications per group per year is more the rule than an exception. In this way the output of physics in the Netherlands is more than three thousand scientific publications per year, enough for the eleventh position on the list of most productive countries in the world of physics. Nobody can handle this scientific tsunami, except Elsevier Science Publishers, which have shown record high revenues and profits. The libraries are full of unread scientific journal articles. No wonder a Science Citation Index has been introduced in which the value of all scientific work may be estimated. The average number of citations the computer registers is 1, so on average all scientific work only gets cited once, but because the good work is cited much more often there is a lot of research not worthwhile to cite even once. The Netherlands scores high on the citation index, scientifically we belong to the G7. Although this score is reason for our government to be proud, it is as much a demonstration of decadence, because scoring in science seems as important as it is on the soccer field.

Today physicists form teams with a captain (group leader), a coach (lab director), and a sponsor (the funding agency). Together they race to their goal in competition with similar teams. Those who arrive first are honoured and the rewards are high. Those who cannot follow, despite all sorts of dirty tricks, will try to have the number one disqualified. If that is not successful either, they are ambitious and smart enough to sidetrack, and have their publicity machine draw their own finish line over which they will rush under self-made applause. For who will forgo the bonuses: public recognition, a good salary, a whole team of assistants and 'soigneurs', and of course a new sponsor for the next race. The industrialized physics after WW II has lost track. It is not about 'atoms for peace' anymore (if it has ever been about that).

What then is it about?

In elementary physics the theoreticians are in the lead. They have the guts to promise their Grand Unified Theory, the ultimate theory that encompasses everything. These theoreticians predict one elementary particle after the other. For the experimentalist in this field what remains is merely the unrewarding task of finding the already predicted particles. No wonder they have made a race out of it. If you go to Geneva to take a look at the largest particle accelerator in the world, you will be duly impressed by this masterpiece of modern technology. Hundreds of engineers have worked on this accelerator for years and it has established the European hegemony over the American and Japanese particle physicists, also for the near future. But of course the American physicists have now asked Congress an amount of 8 billion dollars, merely to build an even bigger accelerator. They also promise to find the 'Higgs boson' that theorists have predicted to exist and so claiming the first position in this prestigious field of physics for America. Particle physics is a daughter of nuclear physics who has taken over and swallowed her mother. The nuclear physicists could have secured themselves had they not lost contact with nuclear energy. Now after forty years both fields are at a dead end. The atomic nucleus has literally been shattered to pieces and our hopes for solving the world energy problem have not been fulfilled. Not either by those who work on fusion instead of fission. For more than forty years fusion physicists have been promising a sustainable and clean energy source, but the only thing that sustains so far is that this source is still forty years away. No one likes to admit that openly, for fear of losing their subsidy (in Europe alone more than a half billion Euros per year).

In my own field, atomic and molecular physics, two symptoms of decadence are easily recognized. First, it is dominated by military money, both in the US and the former USSR, for the development of laser weapons. Second, we move from one 'gee whiz' to the next. We are all members of the 'mutual admiration society'.

Statistical physics is the field of the great Van der Waals who gave our country its leading position in this world. Behind heavily subsidized dykes the physicists of this polder have continued to pump their lowlands dry as dust for decades. Everything that could be measured was measured, independent of anybody's interest. Today the methods and techniques from statistical physics are applicable to the study of complex systems such as polymers and liquid crystals. Until now physicists did not want to be bothered by such 'chemistry'. But that will change since the Nobel Prize for Physics was awarded in this field.

Solid-state physics is one of the most lively and biggest in modern physics, thanks to research labs of AT&T, IBM and Philips. There is a long list of magnificent discoveries, and completely new fields have been created, such as surface science and nanoscience and technology. The investigations have lead to rows of products that have already been brought on the market. Nevertheless, decadence also lies in wait here, as was clear when high temperature superconductivity was discovered. Literally thousands of physicists jumped on the bandwagon. What they were doing before, nobody knows, but apparently nothing very important as they could switch overnight to high-Tc superconductivity.

Since the companies in microelectronics have not been doing so well recently and their researchers have been asked to contribute more to the technology of the company, the scientists complain with indignation about the destruction of their oh so fundamental research, which apparently is not so fundamental for the company's survival. They have not learned yet that inside almost every technological problem there hides some interesting physics, if you only dig deep enough.

The bio- and medical physicists suffer from a minority complex. Their fellow physicists do not take this field seriously because it is not considered fundamental science and much too empirical.

How did it get this far?

During the 'fin de siècle' decadence in physics was not noticeable at all. At the turn of the nineteenth century physics was in one of the most exciting times in its history. There was a heated debate between the atomists and those who did not believe at all in atoms. At first the latter dominated because the atom was much too small to be made visible at that time. The atomic theory, however, could explain many different observations at once, both in the chemistry and physics of those days. Based on the atomic theory at the beginning of the twentieth century it became clear what the origin was of the laws of Boyle and Gay-Lussac; what the number of Avogadro precisely meant; how Mendeleyev's periodic elements should be interpreted; why the theorem of Van 't Hoff was valid exactly; how the specific heat of gasses should be explained; what Brownian motion meant; what this has to do with Smoluchowski's theory of fluctuations; how Raleigh's refraction of light should be interpreted; the absorption laws of Stefan and Boltzmann and black-body radiation; the x-rays; Planck's constant; and the photoelectric effect.

Phenomena that were not well understood until then, and which at first sight did not at all seem connected, turned out to be explicable on the atomic level. The euphoria over this discovery was enormous and has dominated twentieth-century physics. Albert Einstein, the greatest physicist of our time, was misled by the triumph of atomic physics. He was closely connected with the reduction of macroscopic phenomena to simple atomic physics. It must have been such a revelation that he considered reductionism the divine task of the physicists. Einstein devoted the rest of his life to look for the deepest level on which physical phenomena could be brought together, a level where the great formula with which God created the universe should be found.

The consequences for physics have been disastrous. Because of Einstein's search for this deepest level, every physicist thought that real fundamental research on matter should be done in this reductionist way. All applied science was considered inferior to a higher cause: the search for the scientist's stone. In vain, for we now know that inside the atom an infinite world of physical phenomena may be observed. Einstein's ideal is as a retracting horizon: every time you think you have arrived at the deepest level you discover a whole new physical world. The foundations are not to be found. The fundamental task all physicists seem to have inherited from the great Einstein, and which has monopolized twentieth-century physics, has turned out to be a mission impossible. Physics is in crisis; a whole new philosophy of physics has to be invented for Einstein's is inadequate. In absence of the proper philosophy the physicists of the twentieth century have lost track; thus decadence prevails.

1999

Physics and faith

Veltman said he was so late in receiving the Nobel Prize because he was not as good in PR as some of his fellow physicists. Most likely he was thinking of Steven Weinberg who became world famous well before he received the Nobel Prize in 1979; for elementary particle physics based on the mathematical method of Van 't Hooft and Veltman. Two years before, in 1977, Steven Weinberg published *The First Three Minutes*, in which he applied his own research to the Big Bang theory. It became a best-seller because the author promised his book was going to replace the Bible book of Genesis.

After reading *The First Three Minutes* I wrote to Steven Weinberg asking him to explain to me once more how on earth it was possible that more particles than anti-particles were created in the Big Bang. I never received an answer, not even in the articles he regularly publishes in *The New York Review of Books*. Weinberg likes to make firm statements with much ado and a lot of PR, but he is not really interested in a dialogue. In *The New York Review* of October 1999 he writes: 'I am in favour of a dialogue between science and religion, but not a constructive dialogue.'

In the article *A Designer Universe?* Weinberg raises the question whether there are signs showing evolution evolves according to a plan. One would expect that someone who thinks he is able to describe the development of the universe in the first three minutes of its existence using the laws of physics would answer this question in the affirmative. In the quest for the blueprint of creation, Weinberg does not come up with his standard Theory of Everything, but with a counterstatement: 'Show me the passport of the creator.' Thereupon he uses the well-known arguments of the past three hundred years. If the creator would care about us in any way, there should not be misery in the world. Unquestionably, there is a lot of misery, which makes it impossible for the world to be the stage of the Almighty. Consequently, his actions cannot be the explanation for the miracles we observe either. Weinberg does not believe in miracles, even less than in cold fusion. Ever since quantum mechanics, miracles have gone from this world. When we will be able to understand quantum mechanics he does not know yet, but it will only be a matter of time.

Apparently Weinberg does not find it miraculous that the mathematics of 't Hooft and Veltman is the language of science and equally applicable to experiments with particle accelerators as to a description of the Big Bang. That the fundamental constants of nature under certain conditions have exactly the proper value to make a universe as we know it possible, including life on earth, Weinberg waves away with a rather strange argument. Probably there was not one Big Bang but infinitely many and all those other trials did not come to anything just because in all those other Big Bangs the correct numbers did not show up. Nature is like a lottery with many blanks. Survival of the fittest would apply to Big Bangs, fundamental constants and universes.

'The more the universe seems comprehensible, the more it also seems pointless,' says Weinberg, and also: 'One of the accomplishments of science, if it is not to make it impossible for us to believe, then it is that science makes it possible for us not to believe.' That is why he thinks a constructive dialogue between religion and science is impossible.

I refuse to believe Weinberg. Religion and science, faith and curiosity both play a part in our survival, or else they would not have emerged from evolution. They are too costly not to have a function. Thanks to our curiosity we understand nature increasingly better and organize our life such that it becomes more and more comfortable. We also make serious mistakes in the process, serious to a degree that it might become fatal, but at least we realize that, we are warned and we can create a strategy to survive. Thanks to science. We can also learn from religion, even though much of it is not only incorrect but in the eyes of some even immoral. The message of religion is that we have to believe. The scientist also needs faith: the faith that it is worthwhile to be curious, to do science, even though you sometimes do not see the point of it. Although we do not know where it will lead to, our life as a scientist is worthwhile to be lived. We do not know that for sure, but we have to keep faith.

1999

Theories of everything

Suppose we could rewind the film of evolution all the way to its beginning, to the Big Bang, and play it all over again, would it arrive at the same point of today? Will we humans appear on the screen, in a form we will recognize? It does not have to happen all at the same time and in exactly the same way; it may well be somewhat sooner or later and it may also proceed in a slightly different order. The important thing is that the evolution goes according to the laws of nature with the inevitable result of the state of things today.

If the answer is yes, then everything evolves according to clockwork, through which not only the past but also the future is fixed. Then everything is predetermined, not only the evolution of the universe but also the history of the earth and of humans, according to laws that we cannot influence. Is there still room for free will? And responsibility?

If the answer is no, if the film of the evolution is replayed but evolves in a very different way, does this mean that everything is haphazard?

According to the palaeontologists the impact of a huge meteorite terminated the life of the dinosaur on earth. Only after their extermination did humans stand a chance. At the time of the dinosaurs only mammals existed that were small enough to hide. Through the haphazard encounter of our earth and a meteorite so much dust was blown up that almost all animals and certainly the large dinosaurs became extinct. Only after this mass destruction in 'Jurassic Park' could humans emerge in the evolution of earth. Thanks to a magnificent accident.

Nevertheless, we humans are in the need of a Theory of Everything, a Creation Story, a 'Big History'. We do not only want to know how the whole universe came into being, but also why. Most preferably we would like a comprehensive explanation, a few basic data and a formula that agrees with the history of the earth, the origin of life and of humans. Laws with which not only the past, but also the future, can be predicted.

Intuitively it looks like the world consists of elementary particles, that everything is built up from these building blocks, and all phenomena in the universe in principle can be explained by the properties of these building blocks and how they interact. If that were true and if we would know the elementary particles and their interactions existed, for sure we would have a Theory of Everything. This reductionism is based on the notion that things and phenomena at a certain level can be reduced to an elementary building block at a level below. We humans are reduced to neurology, genetics and biology. These are reduced to biochemistry, biochemistry to the behaviour of molecules, molecules to atoms, and atoms to elementary particles. This reductionism has been very useful and important as a research programme, particularly in cosmology and evolution biology, but is there any reason to assume that reductionism to the lowest level is at all possible?

Physicists of the nineteenth century thought their field was almost completely discovered. In 1820 Pierre Simon de Laplace wrote about an intelligent being with infinite mathematical capabilities, who at a certain instant got to know the exact position and motion of all particles in the universe. Such a superhuman being with these data and the laws of classical mechanics would be able to calculate the future of the universe in great detail and at any instant.

Chaos theory, however, tells us that the future of three or more particles, which attract or repel each other, is in principle unpredictable. You may think that we know the motion of the earth and the planets in the solar system so well that if we launch a rocket it will arrive at Venus or Saturn all by itself. In practice, NASA keeps measuring the position of its satellite and steering rockets are used to correct the trajectory and prevent chaotic behaviour. It is the only way for the satellite to reach another planet.

Heisenberg discovered that we cannot accurately know both the position and the velocity of a particle. If we measure the velocity with any accuracy, it takes a certain trajectory to do it. Therefore the exact position of the particle is not determined, but smeared out over the trajectory in which we measure the velocity. The uncertainty relation of Heisenberg is at the basis of modern physics.

'God does not play dice!' Einstein desperately cried, but atomic physics has shown that pure chance plays an important role in nature. Today we can do experiments with a single atom. If it is irradiated with laser light of the right colour, then the atom will start to radiate light in all directions. We see the light from a single atom; under proper conditions indeed visible to the naked eye. If we choose an atom that can store the laser light temporarily, instead of radiating immediately, then we see the spot of light switch on and off at random. It is as if an invisible hand operates the switch. When this invisible hand switches is not determined by anything or anybody. It is pure chance. The difference in duration of the dark periods is most literally an effect without cause. Here the strangeness of quantum mechanics is visible to the naked eye. With quantum theory the time distribution of light and dark is calculated with great accuracy, but the instant the spot of light turns black is completely random. It is always the same atom that always under the same circumstances absorbs and radiates light. Under identical conditions the atom shows different behaviour and the time differences do not have a cause. God appears to play dice indeed.

What abut those dice? Can everything be reduced to the behaviour, be it chaotic or unpredictable, of building blocks? Is nature built up from, to be reduced to, some elementary particles? To answer this question, let us take the eldest and most famous elementary particle by way of example: the electron.

The electron is a point particle. This means that it does not extend in space, it is infinitely small. Yet it has mass. How is it possible for a particle to have mass but no size? Einstein has shown the equivalence of mass and energy: $E = mc^2$. The electron represents an amount of energy. If the electron meets its antiparticle, the positron, both disappear and only a flash of light remains.

To explain the behaviour of the electron it had to be assumed that it is not a particle but a wave. Davisson and Germer proved the wave nature of the electron. By scattering electrons of a nickel crystal, they observed the same interference pattern as for the diffraction of x-rays. In a recent collaboration between Philips and the Technical University Delft it was discovered that transport of electrons through a narrow gate is quantized. The current through the gate jumps with whole numbers of electrons as the opening of the gate is enlarged to one, two, three, etc. wave numbers of the electron.

In the one experiment the electron behaves like a wave, in the other like a particle, but the electron is neither a particle nor a wave. It is an 'electron', and the word stands for all our experiences in experiments with electrons. The same is true for other so-called elementary building blocks of nature, such as quarks or strings.

During the Big Bang matter was made in the form of quarks and antiquarks, at least that is what cosmology says. After that the film of evolution followed its course, with the well-known result. In principle, during the Big Bang equal amounts of matter and antimatter must have been created. It is strange that in the universe antimatter from the Big Bang has not been found. Cosmology stated that already soon after the Big Bang, after 10exp-12 seconds, an asymmetry between matter and antimatter was formed. So shortly after the Big Bang the density of matter was so high that radiation could not escape. Therefore the theory cannot be verified. If we rewind the film of the evolution we have a problem, for it turns out that the beginning is absent.

Reduction as a research programme may have been successful, the idea that complex phenomena in nature at a certain level can be explained by building blocks one level down, does not hold. Not only because the lowest level cannot be reached, just like a retracting horizon. Reductionism ignores the working of the higher complex level on the lower level, which is why those building blocks on the higher level behave differently than below. In collision with a nickel crystal electrons behave differently than in a vacuum. The collective motion of atoms is essential for the working of a laser and differs fundamentally from the behaviour of single atoms.

A star is indeed a bunch of atoms, but through gravity inside the star atoms are brought into nuclear fusion and new heavy atoms are created. So a star is a process rather than a collection of loose atoms.

In principle, the ocean consists of uncountable numbers of droplets of water, but no oceanographer would think of explaining the behaviour of oceans based on the properties of water droplets. A single droplet contains 10exp23 molecules of water, which is more than all water droplets in all the oceans together. Why should the properties of water be deductible from the water molecule? Indeed, water molecules in droplets have a completely different structure than a single water molecule in a vacuum. The sum is more than its parts.

Although physics of the nineteenth century in principle has demonstrated the impossibility of reductionism, the Theory of Everything remains the holy grail of our field. Because of the separation of different fields of science one applies reductionism within one's own borders.

– Cosmology reduces the universe to elementary particles and the Big Bang
– Biology reduces life to metabolism and organisms to cells
– Darwinism reduces evolution to natural selection
– Genetics reduces humans to genes
– Neurology reduces brains to neurons

So, there is not just one Theory of Everything but there are Theories of Everything. What they share are the promises and limitations of scientific descriptions.

Sigmund Freud noticed that great discoveries in science always put humans back into a more humble place in the universe. Since Copernicus we are no longer the centre of the universe. Since Darwin we descend from apes. Since Freud himself we know how much we are influenced by our own subconsciousness instead of reason. In the Scientific American of November 1994 Stephen J. Gould put humans in an even more humble position on earth. In a replay of the film of the evolution, according to Gould, it is most unlikely humans will reappear. Only if the evolution is repeated a great many times, do humans stand a chance.

If we only result from a magnificent accident, and our search for a comprehensive explanation is like a journey to a retracting horizon, what then should we be looking for? Why that longing for a Theory of Everything, a Creation Story, a Big History? Most of all we would like a comprehensive explanation that agrees with the history of the earth; laws with which not only the past but also the future may be predicted. That is the function of the Theories of Everything.

From dead matter life has emerged. Evolution produced, be it by accident, humans with their brains and their consciousness. With both we can to some degree control our own evolution, just as NASA does with its satellites on the way through the solar system; continuously measuring position and steering, to avoid chaotic behaviour. Not only nice to know, also the need to know. Our brains make scientific research possible and the creation of Theories of Everything. Our consciousness helps us as responsible citizens to establish a viable society, in the interest of ourselves and thus of our survival. Why science? For survival!

1994

*To colleagues and friends, for decades of Fun, Utilization, Theories of everything and **Survival**. Thank you*

A mind's eye

Good scientists ask good questions. They do not waste time on problems that have been solved already. Neither are they seduced by the romantics of unanswerable questions. Good researchers have a good eye. They know which questions they can solve. For that they have developed a mind's eye.

Our scientific knowledge may be viewed as an extensive network. Good scientists know how to find the knots, can isolate them and get the main threads in hand to untie them so that the leads to the surroundings become clear. It is not an art to untie knots that are no longer tangled. It is not productive to work on Gordian knots, nor in areas where nothing has been untangled yet. Good scientists add something new to the network that we already know.

Most of the time it is only an irrelevant little knot that they untangle. Sometimes such a knot is at the border of a whole new field and suddenly our knowledge is hugely increased. Spectroscopy sprang from atomic physics in this way and from this came nuclear physics and then elementary particle physics. Sometimes the developments have been going into one and the same direction for so long that it is time for different knots to be untangled. Perhaps that is why in physics after the study of the smallest (elementary particles) and the study of the largest (the Big Bang) we now see a revival of the most complex. The explanation of the behaviour of macromolecules, such as polymers, liquid crystals and colloids has been experiencing a spectacular development, since the researchers dared to ignore the detailed molecular structure and focus on other parameters such as global form and flexibility. Thus one has grasped the leads to untangle the knots in complex liquids.

Sometimes the changes in direction are systematic and occur gradually, sometimes suddenly. Think of the discovery of superconductivity at high temperatures. For eighty years hardly anything happens and most researchers have left the field already, then someone has the idea to try oxides instead of metals and all of a sudden progress is enormous.

How does a new idea come about? The creative moment of one

researcher, the moment his mind's eye suddenly sees how it should be, how does that work? For every scientist who has at one time been able to ask a good question and to his surprise has been answered by Mother Nature, it is a burning question where that good question came from. For a long time I thought that the question on the origin of creativity was a fascinating but unanswerable question, just as all the other unanswerable questions (on the origin of the Big Bang or of life). Our own brains cannot possibly understand themselves, can they? But from the book of biologist and writer Dick Hillenius (*De hersens een eierzeef*) I read: 'Just as I can look at my own eyes in the mirror, so I can think with my brains about my own brains.'

Dick thought an idea the most important thing that can happen to you from time to time. 'Of course not a Platonic idea – a from unwilling senses necessarily superficial image of reality – but, on the contrary, in a flash created through the collision of alert and diverse senses, to see connections of things that have never been connected before.' Dick Hillenius thought that acquiring the idea goes as follows: 'As long as you possibly can, have the senses gulp in the data of reality until the built-in pattern comes to light.' The built-in patterns are the result of 3,000 million years of evolution of life on earth that has led to our brains with which we just have to do it. 'This may also explain why sometimes good ideas are formed on no, too little or even wrong facts. We are not a passive camera obscura. The images that reach us are confronted with everything we have learned and with everything that has become input during three billion years of selection of knowledge, of program, into our computers.'

Consciousness belongs to the physical reality, it does not hover as a soul above us. We may influence that physical reality, so our brains must follow the same laws of physics; they are a necessary part of it. Our soul cannot steer our body without energy, it is not a perpetuum mobile. Both body and soul must follow the law of conservation of energy. Our soul is not something different or higher, but what is it then, what is our consciousness? Hillenius compares its functioning to that of the computer. Both process signals and store them. Most computers are programmed such that they fulfil many different functions: texts, calculus, drawing, games, etc. When making different choices it seems we are working on different machines: typewriter, calculator, drawing board or game machine. Yet we are busy on the very same machine, but in different parts of its programme. You could compare our consciousness to such a programme, but then in our brains. The programming allows our brain cells many different functions. They are partly dependent on the hardware which is primarily genetically determined and partly also on the software, the different menus in the programme, which are primarily learned. Just as a programme may be copied from one computer to another, it is possible to transfer the software of our brains to those of someone else. In this sense our consciousness in principle remains after death, in our children, in our colleagues, in the generations after us whom we have helped raise.

Back to the still unanswered question: How does a good idea come about? Some scientists prefer to use the computer solely as an administrative machine, answering questions of the sort: What is the material with the largest strength, or the strongest magnetic field, or the hardest surface? Such 'expert-systems' know all you always wanted to know but were afraid to ask, because you did not want to look stupid. These computers have superior memories: they know exactly what does not have to be investigated. They are masters in answering questions that have been answered already. They are excellent search machines, useful for a quiz, but they never have a good idea.

Computer simulations are a completely different way of using the computer. In principle, we can put any model-world we like into the computer. In practice this means we think of a model of the system we want to study: for instance a flowing liquid, or the atmosphere of the earth, or even the whole universe. After that we tell the computer what the relevant laws of nature look like. Not the 'constitution', but the law that is applicable to the phenomena under investigation. Then we tell the computer to go ahead: these are the rules of the game, we think, now show us what is going to happen. And after that we just sit back and watch what happens to our creation. One can imagine that in the brains of a scientist it also works more or less in this way: to accurately read impressions of the senses, to digest them as well as possible, to keep memories and experiences awake, to be aware of everything that has made life possible in the 3,000 million years of the history of earth. To model all that and then to look at your brains with your brains, does this not give those scientists a mind's eye?

1992

The mother of all knowledge

If I would have to choose between a travel report of NASA's search for extra-terrestrial life, or a description of the most violent collision between elementary particles in Geneva, where CERN scientists are searching for the origin of mass, or a report on the economy of the Human Genome project, or an essay about the influence of the evolution theory on our time, then I would choose Darwin. Why should a theory, which is already one and a half centuries old, still be attractive today?

In the July 2000 issue of the Scientific American biologist Ernst Mayer argues that the evolution theory has changed our worldview more than any other science. I do not think the evolutionist from Harvard exaggerates and I am even convinced that this change in worldview is still in full swing. To put it even stronger: I believe that in contrast to the physicists the biologists already have the ultimate Theory of Everything, the mother of all knowledge.

Ernst Mayer gives seven arguments.

1. Evolution biology is natural history, an area of knowledge to which both the humanities and the sciences contribute. Here C.P. Snow's 'Two Cultures' work together instead of ignoring each other or competing.
2. Natural history consists of concepts, scenarios and models, instead of laws, rules and experiments such as in the natural sciences. Yet these concepts and models are verified by means of historical and biological data.
3. The evolution theory explains a world without a creator, a development of nature without supranatural forces: a development determined by arbitrariness (variation) and preference (natural selection).
4. No two of the six billion people on earth are the same. Each of us is a unique product of random variations and natural selection, making a genetic base for racism impossible in principle.
5. Natural selection leads to ethical behaviour in social groups. Altruism improves chances for survival of the group and ethologists have already demonstrated this in several kinds of species.

6. The evolution of the world does not develop according to a plan. If the film of the evolution would be rewound and played again, it is all but certain that humans would appear again.
7. The future developments of life on earth are not only determined by physical or chemical forces, or by mechanistic laws. The future depends on purely arbitrary events and on choices we ourselves can make freely.

Looking at Mayer's list of arguments you have to admit that one can speak of a synthesis of humanities and sciences indeed. Extended with dimensions as arbitrariness and freedom of choice, our thinking about nature is much more far-reaching than before Darwin's time. And that had and still has immense consequences for our worldview and self-image: evolution meant nothing less than revolution in the areas such as ethics and religion and I believe that this change will not come to a standstill as long as some fundamental questions remain unanswered.

The evolution theory created a clear distinction between dead and living matter. The first adheres to the laws of physics and chemistry and therefore the future is predetermined in principle. The second, living matter, evolves by changes in circumstances, variation and natural selection, in which the mutation that is best adapted to the changed circumstances will survive, whereas the others will end up at a dead-end. But one mystery remains: how is it possible that living matter is created from dead matter?

As far as we know the universe is fifteen billion years old. The earth was created pretty much at the same time as the sun and the other planets and that happened some five billion years ago. The first forms of life date back to three and a half billion years ago. According to evolution theory increasingly more complex forms of life have come into being, right up until such complex organisms as humans. Perhaps it will become possible for taxonomy and genetics to put the entire evolution with all its branches on a timeline, but up till now there is still something that is gnawing. For is it at all possible, to evolve in only three and a half billion years, from a single cell into our recent, huge diversity of complex organisms? Has there been time enough to produce us merely by variation and natural selection?

The mechanism, which is responsible for the creation of such complex organisms as humans, with brains and behaviour and culture, must continue to be active today. How do we notice that? Is our mental evolution, our political and social development, also the product of variation and natural selection? In society we experiment continuously with new forms of cohabitation. Is the mechanism of natural selection also active here, making it possible for the one experiment to be successful and the other to fail?

In evolution theory the most fundamental questions of this day and age come forward. This makes it the Theory of Everything, the mother of all knowledge.

2000

A matter of civilization

When we look at the earth from space, do we see a ball-shaped candle half burned? Or a spaceship draining its exhaust into its own cockpit? Are we wilfully and knowingly making the climate on earth uninhabitable for generations to come? Have we started a worldwide genocide since the industrial revolution?

Environmentalists are not aglow with optimism. When the earth rose at the horizon of the moon – one of the most impressive images from the twentieth century – we saw for the first time the sparkling white cloud trails over the continents, oceans and ice caps, the delicate variegations of pale blue, green and yellow hues against that pitch black background with glittering stars. We saw our planet alone in an infinite universe. The environmentalists turned it into a stereotype, their advertising message: ball-shaped candle slowly but steadily burning up.

One way of observing the development of humans is measuring their productivity. By using 'horsepower' of his cattle the farmer increased his productivity by a factor ten. The discovery of hydropower meant another factor six and the steam engine produced another order of magnitude. The use of motorized vehicles meant a huge reduction in travelling time, almost a factor hundred compared to the horse, and on top of that it increased the transport capacity to the market. The direct availability of abundant and cheap energy provided many with unknown comfort, mobility and productivity.

Since the first oil crisis we have known that the world has enormous supplies of fossil energy. The picture of the magnitude of the reserves has changed drastically during recent decades. During the 1970s and the beginning of the 1980s we knew for sure the reserves were rapidly diminishing. In the meantime large reserves have been found with the help of new winning and exploration techniques and supplies can be utilized better and more cheaply. The commercially and technically procurable reserves of oil, gas and especially coal, for the next fifty years are twice as large as the cumulative demand of the same period, assuming a moderate growth of the world population and an economic growth of two or three per cent per year. The reserve, which cannot yet be won commercially or technically, is another factor five times the demand and then we have not even counted the clathrates, natural gases which we find in enclosures in deep-lying sea bottoms. If it becomes possible to harvest those, it is to be expected that the world will have enough fossil fuel for many more centuries to come. Even though it sometimes seems as if our globe is ablaze, our candle has not yet burned up.

The admission to and use of large quantities of energy are not equally divided among the countries of the world. In the US the use of energy per capita is thirty times the use in Middle Africa. In the developing countries a minority has access to modern forms of energy; the majority is still dependent on firewood, which is collected daily. Two billion people, one third of the world population, are not connected to the grid. These enormous inequalities cause social, economic and political tensions all over the world.

For the time being there is more than enough coal, oil and natural gas to fulfil the strongly increasing demands, but if we continue to use fossil fuels for our energy needs, it will prove to be detrimental to our health, the environment and the climate. Research on climate change shows that already one-third of all carbon dioxide in the atmosphere is man-made, by burning coal, oil and to a lesser degree gas. If in the coming century the world population will grow to twelve billion people and the energy supply remains dependent on fossil fuels the emission of carbon dioxide and other greenhouse gases will increase from five to thirty billion tons per year. They stay in our atmosphere and transmit sunlight, but not the heat radiation from the earth, causing an increasingly warmer climate. Apart from this enhanced greenhouse effect, there is also acidification of the environment and ultrasmall dust particles get into the air with detrimental effects on our health and that of animals and plants. The shock of recognition which hit us when we saw ourselves from space, was not only due to the realization that our supplies will not last forever, that there are limits to our growth, but also to the fact that we are responsible for our biosphere.

Climatologists, atmosphere chemists, oceanographers, natural scientists, biologists, geologists and sociologists cooperate in large research projects in order to try and understand how the worldwide ecosystem works and are learning to predict the future developments of the climate (IPPC). Their scenario studies show that the average increase in temperature on earth, as a result of the enhanced greenhouse effect, can be as much as half a degree now up until fourteen degrees in a hundred years. The significance of this research can hardly be overestimated: an increase of fourteen degrees is a factor seven more than any natural temperature variation on earth since the last ice age. Moreover, the climate effects are unequally spread over the earth so that local variations can make life impossible very soon. Are recent catastrophes in Bangladesh, Venezuela and here in Europe preliminary signs of this?

Climate change induced hundred and sixty countries at the world conferences in Rio and in Kyoto to agree on a reduction of the emission of greenhouse gases. As a result, the European Community has translated this into agreements with their member states. Thus it happened that the government of our little country recently published a Report on the Execution of Climate Policy, in which it is explained how we can fulfil our Kyoto target to reduce the emission of greenhouse gasses in 2010 by six per cent, compared to 1990.

Some wave climate management away with scornful laughter, and not only in the Netherlands. In the first place there are those who believe in Gaia. They see Mother Earth as a superorganism, which sustains life by itself and sees to it that there is a comfortable climate. Whereas the atmosphere of our neighbouring planets, Venus and Mars, mainly consist of carbon dioxide, the air we breathe is a mixture of oxygen, nitrogen and water vapour, with traces of carbon dioxide, methane and hydrogen in a dynamic equilibrium. The gases on earth react with each other and with the surface of the earth in a combination of biological and geological processes keeping the equilibrium, and consequently life itself, in balance. If the earth were dead, the atmosphere would consist of a composition equal to the one of Venus or Mars. The Gaia believers point out that life on earth has developed for billions of years, in spite of dramatic changes that have happened to the composition of the atmosphere. They suggest human influence on the climate will not be so bad. They do not take into account, however, that the climate changes which have occurred in the past would be catastrophic when placed in our time, not for all forms of life but at least for some considerable part of the world population.

The second opposition against climate management comes from some politicians and lobbyists, the self-interested, who are busy with tactics of delay. Thus in the government policy conditions are added before ratification of Kyoto. The signing by the US and Japan seems a reasonable demand because they contribute most to the emission of greenhouse gases. But they in turn demand that the developing nations must sign the Kyoto targets. And the developing countries do not want to cooperate as long as they use less energy per capita than the population of industrialized countries. We can hardly deny China and India the right to develop in the same way as we have done, but if they continue to burn coal it will undoubtedly be disastrous for our climate. That is why the rich countries wish to have the possibility to implement climate targets abroad. The climate problem is a worldwide problem and it is much cheaper to realize energy saving methods in countries with a weak infrastructure, such as Eastern Europe and developing countries. But the developing countries are afraid that we are going to decide their development and they demand the transfer of modern technology, which we are not willing to give them for free. Another issue in our government policy is the Netherlands' competitive position. If we are the only ones in Europe to stimulate energy savings and renewables by taxation of fossil fuels, then it frustrates our export. So it still takes a lot of negotiating before climate management is a fact.

Of course there are also 'green believers'. A recent survey shows that more than forty per cent of Dutch businesses and households are willing to buy 'green' energy for the current extra price. With surveys such as this a certain bias must be taken into account due to socially desirable answers. But even if only one quarter of the indicated potential is real, it means more than 750,000 Dutch households and more than 50,000 businesses are prepared to buy 'green' energy, a doubling of renewables in our country. So far the market for renewable energy is still small, but with a growth of thirty per cent per year it belongs, together with the ICT business, to the fastest growing businesses in the world.

Is a worldwide entirely sustainable energy household possible? If we do not wish to ruin the climate on earth, if we do not want to suffocate from exhaust gases, we should stop burning the energy supplies of this globe like a candle. The earth is not alone; the sun shines all around it daily. In forty minutes the sun sends just as much energy to the earth as we use up all together in a whole year. In principle, there is more than enough solar energy. The challenge lies in knowing how to make this energy available and affordable for all of us. Solar cells, which transfer sunlight straight into electricity, are still expensive, but in remote areas where there is no grid they are already competitive. In coastal areas with a lot of wind, wind turbines at sea can produce energy. In the North Sea some 40,000 km^2 lies unused and on that area a 'wind farm' could provide electricity for 230 million households. In densely populated areas waste and biomass can be recycled. By gasification synthetic gas can be made, which may be used as transport fuel and as base material for the chemical industry, affordable and avoiding the use of fossil fuels. Is the coming century going to be the century of the sun? Greenpeace thinks so and so does Shell.

In his well known essay *Fire and Civilization* Johan Goudsblom stated that humans through the ages, with constantly increasing specialization and mutual cooperation, have developed ever higher degrees of fire management. The measure of fire control in a society might be seen as a measure of mutual interdependence of the members, a measure of civilization of that society. For Goudsblom the international cooperation in controlled nuclear fusion was the very summit. Nuclear energy does not contribute to the emission of greenhouse gases, but we are left with a radioactive waste problem instead. Is it not as much a matter of fire management and of civilization if we could come to agreement worldwide to stop the emission of greenhouse gases and develop renewable energy? Would we not be at a higher degree of civilization if we would be able to avoid a worldwide genocide?

2001

PS

Whether we like it or not, in the free world the days of nuclear power are almost over. Nobody can build nor insure a nuclear power plant without government guarantee. In a free market economy, however, such guarantee is no longer allowed. That is why for the past decades nuclear power plants have been under construction only in non-democratic countries without a free market economy.

Why science?

1. Consilience

Under severe stress plants produce proline. This amino acid increases the resistance of plants in different circumstances, such as cold, heat, drought, high concentration of salt and UV radiation. Research done by Dr Alia (at Leiden University) shows that the amino acid has positive results because it prevents oxidation damage in plants. The skins of animals and people also contain proline and Dr Alia decided to find out whether the amino acid has a positive effect against aging of mouse skin under UV radiation. This turned out to be the case: the skin of a naked mouse, which is exposed to UV radiation, does not get burned when the animal receives proline in its fodder. In the meantime, a patent has been filed to use proline as a medicine against aging caused by sunlight, certain types of cancer and stress-related illnesses. Thus the botanist Alia has made a discovery, which is of importance for an entirely different area of science.

This innovative discovery is a beautiful example of 'consilience', an expression the British philosopher William Whewell introduced when in 1840 he wrote: 'The Consilience of Induction takes place when an Induction, obtained from one class of facts, coincides with an Induction, obtained from another different class. This consilience is a test of the truth of the Theory in which it occurs.' When several pieces of a scientific puzzle suddenly fall into place, Whewell speaks of 'consilience'. Today two biologists fight for the honour of having rediscovered this word.

The first is Edward O. Wilson, the sociobiologist, who published the book *Consilience: The Unity of Knowledge* in 1998. In this book Whewell is introduced in an effort to unite all sciences. Undoubtedly, Wilson, one of the few scientists who as a writer is recognized by a large public, will have kindled the curiosity of his regular readers with the terminology 'consilience' and with the promise in the subtitle – unity of knowledge – his book has raised expectations in an even more extensive group of scientists. He should have stayed with Whewell's definition of 'consilience', however, after an introduction to Whewell, Wilson seems hardly interested in the British philosopher anymore, and neither in the value of that creative moment when the ideas that come from diverse disciplines suddenly fall into place like a puzzle and offer an insight that convinces just because of the 'Eureka' experience.

Wilson tries to find the unity of science in the scientific method and recommends the reductionism of physics, as if that were the only proper way of doing science, the only way to find 'consilience'. Even now there are still a few diehards in physics who really believe that the entire field of biology can easily be reduced to biochemistry and chemistry is really nothing else but physics, which means that physics remains the base of all natural sciences; but reading this stern reductionism from the pen of a sociobiologist, and seeing him argue that the unity of science can be reached if only all scientists would convert to the naïve reductionism of some physicists, makes Wilson's 'consilience' extremely unbelievable.

In the quest for unity in science, Wilson would have done better to inform himself of the Dutch science historian and philosopher, E.J. Dijksterhuis, who wrote in *De Gids* of 1955: '[...] in the way the different sciences are performed equality dominates over diversity. The aim is always to account for in whatever way established facts, or to connect these with each other. Time and again intuitively a conjecture comes about, how that should be done; or it works through trial and error that a theory is born, an idea conceived, a connection noticed. After that it is ascertained if all known facts are fitting, and finally which consequences may be drawn that lend themselves for further testing. It is the general hypothetic-deductive method, which though most easily demonstrated by using the example of the natural science way of thinking, but which for that reason does not exclusively belong to the natural sciences, as the rules of logic do not exclusively belong to the mathesis, because here their application is most lucid. With regard to this fundamental methodological agreement it is of secondary importance, whether our material of experience has been acquired by observations, by consulting documents or by intuitive comprehension. And the question, whether one should call his or her profession natural sciences or humanities, or whatever, loses its meaning completely.'

Dijksterhuis also recognizes the moment of 'consilience', without using the word, as the flash of insight into a comprehensible coherence, a causal connection, which seems to be able to bring order in the chaos of facts; and the sensitivity for such insight Dijksterhuis calls the scientific talent.

The second biologist, Steven J. Gould, claims to have mentioned Whewell already in 1986. In a more recent book, *The Hedgehog, the Fox and the Magister's Pox*, Gould expresses his disappointment about Wilson, also because of his plea for reductionism that would only widen the gap between physics and all other sciences. According to Gould synthesis between the natural sciences and humanities is possible through 'consilience', for different ideas suddenly coinciding and forming a unity which has been overlooked so far, is an experience that we not only know in the natural sciences but share with the humanities and all other sciences. Moreover, the most interesting scientific questions nowadays are found in the areas where different scholars meet, and where a multidisciplinary collaboration is necessary. In the case of Dr Alia that was biology and medicine.

2. Science as struggle

Without a scientific attitude and method no science, and without 'consilience', without 'Eureka' no progress in science. But that does not mean there is synthesis in the sciences if we can ascertain that all sciences in principle use the same, if not reductionistic method, and that from different disciplines once in a while different ideas suddenly fall into place and 'consilience' is reached. The fantastic flash of enthusiasm, the Eureka effect, which belongs to a creative moment in research, is well known to all researchers from all disciplines. But this does not necessarily mean there is unity in science; on the contrary.

The universities have become multiversities with many faculties, which in turn have been divided into disciplines, professional groups and sections of specialists. It is no longer possible to do innovative research without specializing oneself in a far-reaching way. By increasing efficiency and worldwide competition, it is almost impossible to keep up with scientific production except for the specialists who meet regularly at international and specialized conferences. Thus scientific education and research have fallen into pieces literally and completely.

Science has also become a race, with team leaders, assistants, coaches and sponsors. The prizes are worth it: money for the next race, honour and glory, media attention, popularity with young researchers and sponsors, large transfer bonuses. No wonder some cannot withstand the pressure and violate integrity so that scientific codes of conduct have become mandatory.

Among the different sciences competition has become the rule and collaboration an exception. Since WW II, since the atomic bomb was developed in the Manhattan Project, doing physics has become 'big science' for both governments as well as industry; in laboratories with hundreds of researchers around one or more big facilities that can only be realized and supported in large collaborations. Over time this model has been taken over, freely or not, by the other sciences, therefore we now have Centres of Excellence, Networks of Excellence, Technological Top Institutes, Top Graduate Schools, virtually or for real and pretty much in all disciplines and at all universities at home and abroad. The institutes and graduate schools are continuously in competition with each other, looking for money from the government and business.

The result is a scientific production which is unheard of and that is hard to keep up with for most people. Scientists also depend on science journalists when it comes to translating what is happening in those areas that are not part of their specialism. Under such circumstances who can make a responsible choice and decide on priorities among the different branches of science? Chances are that the hype will win and consequently it is a good idea to create such a hype, together with your colleagues, associated companies, learned journals, news media and politicians. We know them by now: Aids, genomics, nanotechnology, quantum computing, bioinformatics, tissue engineering and most recently Mars again.

Nuclear physics was not the first Big Science (see also Scientific Life), but it was an industrial way of doing science. I remember the early days well, when our budgets grew by 25 per cent per year and justification to society was not needed for the public understood we were working on 'Atoms for Peace', in the very forefront of science and to solve the world energy problem. At least that was what we said, until we found that society did not want our products, nuclear reactors, because we had not foreseen the disasters of Chernobyl and Three Mile Island and had disregarded the problems of nuclear waste. Have we learned anything from our mistakes? Or are we making the same errors today with ICT, biotechnology, and pharma?

3. Resilience

All over the world science is suffering from symptoms of decadence. The universities were not first to notice the problems, but business, and they have taken a different course. After WW II the laboratories of Bell, GE, and IBM in America and Philips, Shell, Unilever and AKZO-Nobel in our country have been able to grow into the largest and most excellent research centres, winning the struggle of science in almost every discipline and sometimes even winning Nobel Prizes. For a very long time their reputation was enough to secure themselves of financial support from the mother company. This has come to an end and large industrial laboratories have dropped out of the race, but have been made dependent on the revenues of the company for which they are working. The result is a poor performance as far as pure science is concerned.

Now it is the turn of the universities. The 'captains of industry' and the politicians demand us 'knowledge workers' to contribute to a 'knowledge economy'. In our country there is a debate in the so-called Innovation Platform with the premier as chairman. The interest of the elite of industry and politics is a good sign. Yet, I am not sure that the discussions are going into the proper direction. They have discovered something: the knowledge paradox. The Netherlands scores fantastically in the science race, measured by productivity worldwide we are part of the G7 and our citation score is even better, as a result of the many publications in the 'high impact' journals *Nature* and *Science*. But we lag behind in economic growth and our scientific and technological innovations have little effect on our economy. The utilization of our knowledge and expertise leaves much to be desired. In the Innovation Platform the same powers are active which have made industrial research dependent on the financial results of businesses.

This raises the question for the universities: Why science? The unity of sciences is also a point of discussion because we should not allow ourselves to be set up against one another with debates about humanities versus natural sciences, between fundamental versus applied sciences, between short versus long-term research, between education versus research. It will not be sufficient, I am afraid, to plead for 'consilience', not in the way Wilson does by canonizing the method of physics, nor the 'consilience, the 'Eureka' of Whewell or Gould.

I believe that we scientists need to rethink the function of science and I find it remarkable that neither biologist, Wilson nor Gould, has looked for the unity of science in the evolution theory, in resilience rather than 'consilience'.

The evolution has not only brought about humans but also their culture, science and technology included. Since the Enlightenment peoples with modern science have larger chances of survival than peoples without; with the result that in a short time almost all peoples from all over the world have availed themselves of science education and research. Thus we do not only speak of a knowledge economy, but of a knowledge society. It does not merely concern innovations for economic growth. It also is so, and ought to be, that the creative moment in science, that new insights into nature and nurture, that 'consilience' contributes to resilience, to sustainable development, to survival. Does the evolution theory not teach us that our culture and consequently our science education and research, our universities have been given a prominent position in society, for the very reason that they contribute to resilience, to sustainable development, to survival? But also that this prominent place can only be safely held for as long as the universities are an adaptation to society, for as soon as they lose this function they are too costly and will be marginalized and eventually even disappear from our culture.

I see the development of science from an evolutionary perspective. In scientific evolution, just as in biology, all sorts of mutations can occur, mutations which are the fruits of the freedom and creativity of the scientist. But selection belongs to scientific evolution too, just as in biology. 'Survival of the fittest'; not only arbitrariness but also convergence. Frans de Ruiter put it this way (*De Gids* Sept. 2003): 'Creating something new of which nobody has been able to foresee its indispensability.' So it is all about creativity and indispensability, about 'consilience' and 'resilience'.

Research is incredibly costly, particularly for the scientist. What is it that makes research worthwhile? Not just the Eureka effect, not just the kick of a new idea, not only the race, the prizes, the honour and the fame. Those have already led to generations of scientists who do not wish to be bothered by hunger, poverty, migration, energy crisis, climate change and environmental waste. Scientists who have been raised to solve the puzzle they created themselves and who do not wish to be troubled by a guilt complex because of the impact of nuclear reactors, weapons, computers, telecommunication or biotechnology on our society.

My conception of the social responsibility of scientists is that they should be aware as much as possible of all signals, factors, forces that influence them and from which they have to make a conscious choice. Which choice, that is every person's own decision. Creative freedom and indispensability! Not only natural sciences also humanities are indispensable, not only science and technology, also beauty and consolation are indispensable. In order to survive. Does this not close the gap between nature and culture? Why science? To survive.

2003

Spinoza's God

'I believe in Spinoza's God, Who reveals Himself in the lawful harmony of the world, not in a God Who concerns Himself with the fate and the doings of mankind.'

EINSTEIN, 1929

Father Minderop was my favourite teacher. With compass and ruler under the arm he trudged into the classroom, his black habit white with chalk dust. I do not remember a schoolbook, except the notebook in which we copied the axioms, definitions and constructions from the blackboard. Father Minderop taught us Euclid by heart; he made drawings on the blackboard and wrote down the explanation. Once in a while he turned around, not to keep order but to encourage us. On our birthdays, he said, we should not believe those aunts: mathematics knots do not exist and you do not need them for this simple and logical subject. Then he turned to the blackboard again and demonstrated how through two points one and only one straight line can go. With compass and ruler he divided angles and segments in two. He constructed equilateral triangles and other objects. He finished his lesson writing down a few propositions for us to prove as homework. Learning words I disliked, honestly I was too lazy to do any rote learning. In mathematics that was not necessary, for you could deduce everything and prove things logically. With Father Minderop we hurried to show our homework. He asked us to come forward to demonstrate our proofs with the whole class as witness. It was most exciting that for almost every proposition there was more than one proof possible. Poor Father Hirsch, our religion teacher, who tried the proofs of God on us, but failed bitterly. I lost my religious conviction, fascinated by the exact, demonstrable world.

My current conviction I do not owe to the clergy but to my parents, my biologist mother and my physicist father. One Sunday my father called me and handed me a copy of his PhD thesis with the following dedication: 'Frans, Science requires love for the truth, disinterested dedication and perseverance. Thus you will contribute to the evolution of humans and of nature and through that to the fulfilment of a Divine mission.' He added that it was time to replace the religious conviction of my childhood for my personal belief. Now I believe in Spinoza's God or Nature.

The *Eureka! Wetenschapsprijs 2002* and the *Gouden Uil 2002* both went to Bas Haring for his book *Kaas & de evolutietheorie*. As far as I am concerned he could also have received the 'Board Game Award 2002' for the invention of a board game with which he explains how evolution works. He called it Abalone and it is played on a board divided in boxes with black and white balls in them. You do not know the rules and you play against the Abalone champion, but you have a hundred boards, on which you always kick one of the white balls in a random direction. If because of that a black ball rolls off the board, the opponent loses one of his balls. Apparently, you have to roll the balls of the opponent off the board, but you do not know for sure. After your turn, almost all your moves appear to have been wrong and those boards disappear from the table. The game continues with the boards on which you haphazardly have made the right move. Upon your next turn the number of boards have been increased to a hundred for each board that was not removed from the table. Therefore you try again a hundred times, but again all boards with wrong moves disappear from the table. A passer-by, who sees you play against the Abalone champion, will get the impression you are doing very well, because he will see only the boards with the right moves. All other boards have disappeared. It looks like you are playing according to some ingenious system, whereas you do not even know the rules. Yet you get the chance all the time to make a hundred moves of which only the right ones survive. In this way Bas Haring illustrates how evolution via random mutations and natural selection works.

According to modern science, Big History, of both living as well as dead matter – created some fifteen billion years ago by the Big Bang – is the product of infinitely many mutations followed by natural selection. Evolution is natural history, an academic discipline to which the natural sciences, the social sciences and humanities contribute. It consists of concepts, scenarios, and models instead of laws, rules and experiments. Yet those concepts and models may be verified using historical and biological data. The history is still far from complete; there are still many *missing links* in this field of science. From cosmology an explanation is expected for the origin and structure of the universe, whereas 90 per cent of all matter and energy according to the astronomers is 'dark matter' and 'dark energy', i.e. completely unknown. From biochemistry an explanation is expected on the origin of life from dead matter, although we do not even know exactly in what sense living matter differs from dead matter. The evolution does not evolve according to a plan; if we would rewind the film of the evolution to the Big Bang and play the film again, it is not likely that we humans will appear again. Yet from molecular biology we expect the whole tree of life as the logical and sequential ordering between all organisms in the evolution should be derived from genetic information. Much is still unknown about the Creation; yet a complete explanation is expected, not from religion but from science.

Before Darwin humans were like passers-by at the Abalone game: they believed in a Creator who guided his creation according to an intelligent design. Due to the sciences little is left of this belief. The conflict between religion and science, which began already with Galilei, has turned into an intense blaze since Darwin, but today it is over, except for some rearguard battles, such as between creationists and evolutionists in the US. Or in the Netherlands where the Minister of Education submits plans to Parliament in which physics has largely disappeared from the secondary school curriculum and religion is mandatory. For our research organization and for our media the conflict is over; together they organize the National Science Quiz, which has become rather popular, despite the fact that it is shown on TV on Christmas Eve. Also for Bas Haring the conflict between religion and science is finished. He says: 'In what kind of God do people believe who believe in both God and the evolution theory?'

This question may be answered in many different ways. Newton and Bacon believed in a personal God who steered their lives. For Pascal things were not so simple: 'If nothing pointed to God's existence, I would become a disbeliever. If proofs of God were to be found everywhere, I would safely and happily believe. But I see too many proofs to deny the existence of God and too few proofs to be entirely convinced.' Even modern scientists recognize this point of view from a man who lived between 1623 and 1662. But with that it is still not clear how they rhyme their religion with the evolution theory. According to Steven J. Gould that is not at all necessary. For him science and religion are two 'non-overlapping magisteria' (NOMA). The one is the 'magisterium' of reason, the other of meaning, or purpose.
And according to Gould these two exist in completely different spheres. Perhaps Casimir meant the same when he wrote about revolutions in physics: '[...] an approximate description of a limited part of the physical phenomena, which form in themselves only a limited part of our human experience.' There are also religious scientists who think you should not take religious texts literally, but as symbolic recommendations on how to lead a virtuous

life. In his book *Can a Darwinian be a Christian?* Michael Ruse answers his own question with a firm 'yes', after he has first explained extensively that a Christian does not have to believe literally in *Genesis*, but has to believe in Christian morality. David Sloan Wilson goes much further in his *Darwin's Cathedral* and discusses a number of different moral religious doctrines: the one of the Balinese water temples, those of the Calvinists in Geneva and their predecessors, the one of the community of Catholic South Korean settlers in Texas, and those of the first Christians in the Roman Empire.

The Dutch philosopher Herman Philips is not having any of this, for him they are merely 'the emperor's clothes'. In his *Atheïstisch Manifest* he deals with the four possible strategies of religious scientists: 1. the faithful who stick to Genesis are guilty of being an 'intellectual ostrich'; 2. The religious/theologian who disclaims any conflict between religion and science by denying overlap (NOMA), commits 'theological suicide'; 3. The religious/theologian who practises theology as a science of religion (according to which the real conviction of people might be described in an empirical way) practises science as someone with a 'disability' – it is better to stick to 'bona fide science'; and finally 4. the religious who explains the traditional texts only in a symbolic way, and take them as moralistic recommendations, 'stops being religious'. Philips cannot escape from embracing atheism. It is remarkable that the philosopher does not have a view of God for the Darwinist.

It is even more remarkable seeing the Darwinists have given belief in God and religion a proper place in the evolution of humans. According to sociologists and historians religion was absent in humanoids and the first humans. But there are archaeological data, from periods when humans started to live together in larger groups, indicating cultures with religion – certainly when they evolved from hunters and gatherers into farmers. Religion stimulated group cohesion and harvest rituals directed by priests helped to deal with abundance and to save for a bad day. In short: there must have been a time when peoples with religion had a higher chance of survival than neighbouring peoples without, and therefore in time all peoples all over the world have become religious. The evolution theory is a universal theory indeed, in which the belief in God and religion has been given a recognized place and function in our cultural history.

Is there no place for this belief anymore since Darwin? Is religion merely for prehistoric times and not for modern men? D.S. Wilson says: 'Theory of God cannot be hostile to God.' But Philips objects to this attitude, 'because she ridicules and misinterprets a series of issues that deserve the serious attention of us scientists. [...] The atheist rejects religious doctrines precisely because they are untrue. [...] In the course of history the function of explanation has split-off and has become science. Therefore religions are in crisis. [...] We better establish social cohesion in our society by other means than by the illusion of religion.' But apparently Philips lacks 'other means' when he discusses the question whether the commandment of love of one's neighbours has to be extended to the whole of humankind. In his *Atheïstisch Manifest* we read: 'Does the commandment of worldwide love of thy neighbours mean that Western countries should share the wealth of eight hundred million people with a world population of six or seven billion? Then everyone will live in poverty and the economic institutions that secure the wealth of the West will be ruined. Or does the commandment mean that Western countries should help the rest of the world to acquire the level of economic development and consumption characteristic of the West? That will result in an ecological catastrophe. In both cases the commandment of universal love of thy neighbours probably means the end of Western culture as we know it.' Philips acknowledges the importance of social cohesion, but dismisses a morality based on religion, whereas it remains unclear what should be the basis of our morality.

Bas Haring has also lost track: 'Nothing indicates that life has emerged with a reason or a goal. We have come about and here we are. There is not even anything universally sacred in the service of which we may view our lives. This does not mean your or my life is meaningless. Life itself can be meaningful, only the origin of life has no goal. And that is after all really very nice also. For it gives us the opportunity to determine our own goals.' This point of view seems to me not entirely harmless in these postmodern times in which his book is so widely read, especially by the young. Is there no a higher goal for Darwinists than to beat the Abalone champion?

Does the nice metaphor of the Abalone game not fail for the social and cultural evolution of humans? We have been passers-by long enough to know the rules of the game and to guess which moves will lead to survival. Is sustainable development not what we have learned from nature and Darwinism? Humans have been dominating evolution; humans can to some extent determine evolution, at least their own evolution. We can choose from the very many random mutations, natural and societal developments that occur to us regularly. Our preference is that which contributes to our survival. Our ethics is survival ethics. We can destroy ourselves, and the earth; but we are conscious about it and we try again and again to find ways to survive; we consider that to be our moral duty. From evolution theory, survival morality follows.

But this also goes along with belief, for why should we necessarily have to survive? That is not only a matter of a biological drive to live but also of belief, the belief that it is worthwhile to survive, the belief in nature, the belief that evolution without a plan will still lead us somewhere? Belief in God?

You have to believe, that holds also, or is perhaps especially true, for the scientist. Why science? To survive. Since the Enlightenment peoples with modern science have a higher chance of surviving than people without, so in a short time all peoples all over the world have started to believe in modern science and technology. For the individual researcher there is also an extra issue; for those who have experienced the Eureka effect that belongs to a creative moment in research, to make a discovery is a divine encounter. Apart from the kick of the discovery, to contribute to sustainable development for me personally has made life worthwhile. That I do not know for sure; it is my belief. I believe in nature and in science. This belief helps me not only to survive, but it also helps me to make moral choices: what to investigate and what not. Synthesis between religion and science is indeed possible and for the researcher I think even necessary. My answer to the question: 'In what kind of God do people believe who believe in both God and the evolution theory?' is: 'Nature, the whole universe, everything that is, including us humans.' So, I believe in Spinoza's God.

2003

Science through the looking glass of literature

Mr Rector Magnificus, ladies and gentlemen,

> '*In this world, the passage of time brings increasing order. Order is the law of nature, the universal trend, the cosmic direction. If time is an arrow, that arrow points toward order. The future is pattern, organization, union, intensification; the past, randomness, confusion, disintegration, dissipation.*'

Much is known about Albert Einstein, but not what he was dreaming when he worked on his special theory of relativity. For Alan Lightman, physicist and writer of fiction, whom I have just quoted, this is an excellent opportunity to conjure up twenty-four theoretical realms of time, in as many fables dreamt of in just as many nights. All are visions that gently probe the essence of time, the adventure of creativity, the glory of possibility, and the beauty of *Einstein's Dreams*. In short stories of three to four pages each, Lightman creates twenty-four wonderful little worlds, worlds which Einstein may well have been analyzing, beautifully described with fascinating details; also unfolding short philosophies as we know them from Einstein's later writings.

In real life the effects of Einstein's relativity are so small that we do not notice them. The fascination of Alan Lightman's fiction is that he makes a link with our own surprising experiences and the enigmatic nature of time. These experiences have led to sayings such as: time flies, at all times, for the time being, in good time, out of time, at the same time, to make time, or to keep time, in no time, etc. For every one of them, Lightman has made up a short story, a dream. From one of *Einstein's Dreams* I have chosen to quote my favourite arrow of time. It is the arrow pointing toward order and as you will hear later, modern science teaches us that this dream will come true.

In 1987, on the occasion of the 150th anniversary of the literary journal *De Gids*, Hendrik Casimir, one of the big shots of twentieth-century science in the Netherlands, asked the following question. Suppose that in a few thousand years' time a future archaeologist would be brave enough to start digging in the then still radioactive ruins of our civilization. And this archaeologist would miraculously find a collection of poems, but no other books or manuscripts. Would he get any idea of our civilization? Casimir put together a collection of poems from his favourite authors and concludes that there is such a big gap between literature and science that the archaeologist would not get a truthful image of our society at all. *'Scientific knowledge is a substantial part of our knowledge, science-based technology is an essential element of our material world. Is this also reflected in our poetry? I do not think so.'* Casimir concludes.

Casimir is not Gerrit Komrij and perhaps it is worthwhile to start a search for modern science and technology in Komrij's collection of Dutch poetry. But why should this search be limited to poetry? Why could the archaeologist not find prose, fiction as well as non-fiction? From the start in 1837 the literary journal *De Gids* has published poetry, short stories and essays. It would be bad if one could only tell from the dates on the cover of this journal in what time its content was produced. Suppose Casimir's archaeologist would miraculously find all bound editions of *De Gids* of the twentieth century in good order, but not any other books, would our civilization and especially our science be reflected in this collection of literature?

To answer this question a bibliography of all articles on natural science in *De Gids* from 1900 to 2000 was produced. It contains as many as 929 articles from 340 authors, together 9000 pages of science, almost 8 per cent of the total output of the journal in the twentieth century. From these Rob Visser and I have chosen 60 essays and also a few poems, which we have put together in a book (*Trots en twijfel: Kopstukken uit de Nederlandse natuurwetenschap van de twintigste eeuw*). The articles selected from the first half of the twentieth century reflect what some historians of science have called the Second Dutch Golden Age, with Nobel Laureates like Van 't Hoff, Van der Waals, Lorentz, Zeeman, and Kamerlingh Onnes. In the second half of the century, the articles are more concerned with the revolutions caused by twentieth-century science. Indeed, relativity, quantum mechanics and the Big Bang have dramatically changed our view of the world whilst the bomb, computers and lasers have radically altered world order. These developments in science are reflected by the non-fiction literature in *De Gids*. However, by limiting ourselves to essays on science from *De Gids* we have not tried our utmost best, for we have not at all done justice to the full spectrum of literature, fiction in particular. Here I want to present a short anthology of science in literature.

My champion is Harry Mulisch because physics and astronomy play an important role in *The Discovery of Heaven*. In heaven they discover that because of the development of modern science humans no longer believe in God but in humans only, therefore the Holy Alliance will be withdrawn. In a fantastic plot, which in human eyes could only be due to pure chance but in reality is guided by the invisible hand from above, the Ten Commandments are returned to heaven. In passing Harry Mulisch makes several 'discoveries' in physics and astronomy worth analyzing.

It is well known that we do not see the Milky Way as it is now but how it was some time ago, the time it takes for light to travel from the stars in the Milky Way to us. Astronomers look deep into the history of the universe. Harry Mulisch turns this around and says that images of historical events on earth are travelling away from us into the dark universe at the speed of light. The little green men on Mars should be able to see Earth as it was a few minutes ago, and to those who live much further away in the universe the history of Earth some millions or even billions of years ago should be visible. Although we would like to rush after those images, time travel will not help, for one cannot travel faster than the speed of light, unless... We could look at the light reflected from objects in the universe! Indeed, Mulisch has the fantastic idea of looking at our own history using the most sensitive telescopes and making images of the light from Earth reflected by interstellar objects back to us! Who would not want to be able to see the history of our planet and its people? Mulisch suggests we should use our most sensitive observatories not so much to study the history of the universe but rather the history of Earth. Alas, hardly any light from Earth gets reflected by interstellar objects – the material density of the universe is too small – but the mere idea is brilliant.

In his *The Discovery of Heaven* Harry Mulisch also describes a radio astronomer who suddenly realizes that the strange signals he received might have come from the very spot in the universe where the Big Bang has taken place. That is why signals from this place show more red shift than from anywhere else in the universe and why these signals at first seemed so incomprehensible. The astronomer has discovered the infinitely small, infinitely dense, place of appearance and disappearance: heaven itself. He will not be able to tell his colleagues about his discovery; for a stone from heaven kills him instantly. As much as this appeals to our imagination, we will never be able to view the Big Bang in this way. Only from a position outside our universe one might observe the Big Bang as Mulisch imagines it. Unfortunately, we are not in such a position because we are inside this place of appearance, we are not standing outside but we are part of the Big Bang and we see the universe expand around us. We cannot possibly view the expanding universe from outside; it is not the place for humans but for God.

Some twentieth-century scientists have taken the place of God and Harry Mulisch believes that it will lead to disaster. I quote:

> 'To the old global disasters are now added the ravaging tidal waves of the new: with their Baconian control of nature, people will finally consume themselves with nuclear power, burn themselves up through the hole they have made in the ozone layer, dissolve in acid rain, roast in the greenhouse effect, crush each other to death because of their numbers, hang themselves on the double helix of DNA, choke in their own Satan's shit.'

This pessimism is typical for modern literature. Mulisch is not alone: quite a few writers are convinced that, in absence of the steering hand from above, disorder and chaos is the universal trend which is due to a fundamental law of nature. However, as we shall see later, this is not the proper perception of what science teaches us.

For a glimpse of relativity I turned to Alan Lightman's short stories, for the Big Bang to Harry Mulisch's fiction, now for quantum mechanics, the third revolution in modern physics, I prefer drama. In his play *Copenhagen* Michael Frayn introduces three characters: Niels Bohr, his wife Margrethe and his colleague Werner Heisenberg; they represent three issues: quantum mechanics, public perception of science and the making of the bomb.

With quantum mechanics, classical nineteenth-century physics came to an end, albeit not abruptly, for it took the two heroes Bohr and Heisenberg three years to make sense of quantum mechanics. Today, to most physicists it still seems strange: how can a particle also behave like a wave; how is probability reconciled with causality; what goes on during a physical experiment before the measurement? In conversations between the three actors Frayn conveys the essence and the strangeness of quantum mechanics, the uncertainty or rather the indeterminacy principle.

'Bohr: It starts with Einstein. He shows that measurement – measurement, on which the whole possibility of science depends – measurement is not an impersonal event that occurs with impartial universality. It's a human act, carried out from a specific point of view in time and space, from the one particular viewpoint of a possible observer. Then, here in Copenhagen in those three years in the mid-twenties we discover that there is no precisely determinable objective universe. That the universe exists only as a series of approximations. Only within the limits determined

by our relationship with it. Only through the understanding lodged inside the human head.'

With these words Frayn intimates that according to Bohr and Heisenberg we will never know what matter is nor what it is made of, but this does not prevent us from using quantum mechanics to properly predict the outcome of experiments. For Bohr, however, understanding of physics meant being able to explain it to his Margrethe.

'Margrethe: Explain it to me? You couldn't even explain it to each other! You went on arguing into the small hours every night! You both got so angry!'

One of the forms of uncertainty touched upon in the play is the uncertainty of human memory, or at any rate of the historical record. Heisenberg's role in WW II becomes the embodiment, the epitome of uncertainty. Has he worked for or has he sabotaged work on Hitler's bomb? That is the question.

'Heisenberg: Most interesting. So interesting that it never even occurred to you. Complementarity, once again. I'm your enemy; I'm also your friend. I'm a danger to mankind; I'm also your guest. I'm a particle; I'm also a wave. We have one set of obligations to the world in general, and we have other sets, never to be reconciled, to our fellow-countrymen, to our neighbours, to our friends, to our family, to our children. We have to go through not two slits at the same time but twenty-two. All we can do is to look afterwards, and see what happened.'

Of course we need textbooks to teach relativity, cosmology and quantum mechanics, but I think together with our students we should also read *Einstein's Dreams*, *The Discovery of Heaven* and Michael Frayn's play *Copenhagen*. By studying science through the looking glass of literature you see the philosophy, history, sociology, ethics and the public perception of modern science.

At the border between literature, science fiction and suspense we also get a view of science of the twenty-first century. Perhaps the most imaginative in this genre is Michael Crichton. I quote:

'He could not have wished a more knowledgeable audience. The Santa Fe Institute had been formed in the mid-1980s by a group of scientists interested in the implications of chaos theory. The scientists came from many fields – physics, economics, biology, computer science. What they had in common was a belief that

the complexity of the world concealed an underlying order which had previously eluded science, and which would be revealed by chaos theory, now known as complexity theory. In the words of one, complexity theory was "the science of the twenty-first century."'

First in his *Jurassic Park* and then in *The Lost World*, from which this quotation is taken, Michael Crichton leaves no doubt about what he considers the science and technology of the future. Just as physics was the science of the twentieth century, life science will be the science of the twenty-first century. Like so many writers Crichton sides with Mulisch: the genetic engineers through their greed and arrogance will convert the Earth into a frightening game park, leading to chaos and the extinction of humans.

To express his concerns about the developments in twentieth-century science and the end of classical physics, Bertrand Russell (in *The ABC of Relativity*) quotes four lines from Lewis Carroll:

' But I was thinking of a plan
To dye one's whiskers green,
And always use so large a fan
That they could not be seen'

The same lines were quoted by Eddington in *The Nature of the Physical World*, but with a larger metaphorical meaning: the habit nature apparently has of forever concealing from us her basic structural plan. In the century since Lewis Carroll a whole library of literature has been created representing the role of scientists both in fiction as well as non-fiction, but during this century the optimism of the Enlightenment has disappeared and has been replaced by postmodern pessimism. The looking glass has stained and darkened considerably, leaving a rather gloomy fragmented and essentially distorted picture.

Modern scientists are literature's least favourite sons. This is the main conclusion of Roslynn Haynes' book *From Faust to Strangelove*, a comprehensive representation of the scientist in Western literature. Drawing on British, American, German, French, Russian, and other examples, Haynes explores the 'persistent folklore of mad doctors of science' and its relation to popular fears of a depersonalized, male-dominated, and socially irresponsible pursuit of knowledge for its own sake. She concludes that very few actual scientists – with the exceptions of Isaac Newton, Marie Curie and Albert Einstein – have contributed to the popular image of the scientist. On the other hand, the fictional characters, such as Dr Faustus, Frankenstein, Moreau, Jekyll, and Dr Strangelove, have been extremely influential in the evolution of the unattractive stereotypes of scientists. Roslynn Haynes argues that this is primarily due to our lack of communication. In her own words:

> 'By failing to discuss with non-scientists what they are doing, scientists not only endanger society but limit themselves and their research in a number of ways. They may fail to perceive directions that would be profitable to their work; they may fail to convince funding bodies that what they are doing has any economic or social value; they may be left with no control over what is done with their research; and they will almost certainly be diminished as people.'

This is all very well and communication is important, but there is more. It is not only the *public* perception of science; it is also the *proper* perception of science that is at stake. In his most recent novel, *State of Fear*, Michael Crichton mixes fiction with a number of graphical representations of scientific results. Is this the ultimate synthesis of science and literature? Crichton's message comes out loud and clear: *the proper perception of science is essential to the scientist and non-scientist alike*. As Crichton shows, the perception of data on climate change depends very much on your cultural setting. Depending on whether you are in the automobile or oil industry, or if you are a member of Greenpeace, whether you live in a wealthy nation below sea level or in one of the developing countries, your perception of CO_2 emission data will differ greatly. On top of this there is the wilful ignoring of scientific data a particular group does not like.

It will have become clear that, since C.P. Snow, something has changed in our culture. The scientific revolution brought about by Einstein, Lorentz, Bohr, Heisenberg and others has changed our society and our worldview as is reflected in our literature also. Are the basic findings of modern science properly perceived in our culture?

In 1956 C.P. Snow lamented in his *Two Cultures*:

> 'A good many times I have been present at gatherings of people who, by standards of the traditional culture, are thought highly educated and who have with considerable gusto been expressing their incredulity at the illiteracy of scientists. Once or twice I have been provoked and have asked the company how many of them could describe the Second Law of Thermodynamics. The response was cold: it was also negative. Yet I was asking something which is about the scientific equivalent of: Have you read a work of Shakespeare's?'

What is it about the Second Law of Thermodynamics that makes it as important as Shakespeare's work?

Most people believe the Second Law to say that in nature there is a tendency toward the maximization of disorder, but that is a dramatic mistake! Since C.P. Snow the Second Law is widely quoted in scientific and non-scientific literature. But actually prior to C.P. Snow, Schrödinger, with his little book *What is Life*, already raises the question: How could life come about, how may order emerge from disorder, if the Second Law of Thermodynamics says that in nature disorder is maximized? No wonder C.P. Snow drew attention to this; it is a fundamental and enigmatic issue indeed. Non-fiction writers such as Richard Dawkins, Stephen J. Gould, Peter Atkins, Ilya Prigogine and Brian Greene have also basically followed Schrödinger. In numerous articles and books they paint a bleak and pessimistic picture of our future. If nature tends to maximize disorder, according to these authors evolution is merely a pointless succession of mutations and natural selections, in the end leading to nothing but randomness, disintegration and chaos. If famous non-fiction writers send out this message, science journalists will copy them and no wonder the fiction of modern authors like Michael Frayn, Harry Mulisch and Michael Crichton is as pessimistic and pointless as it is popular today. Since C.P. Snow the two cultures have united and have found a common ground in postmodern pessimism. In both the sciences and the humanities it is widely believed that, in the absence of Providence, nature's fundamental driving force leads to maximum disorder and chaos; whereas in reality everything in life shows evidence to the contrary.

In contrast to what is commonly thought, the Second Law is not about disorder, and maximizing disorder is not a driving force of nature. The Second Law is about the energy nature requires for itself in order to increase its freedom of movement. The Norwegian scientist Onsager was the first to draw attention to this, already in the days of Schrödinger. More recent laboratory experiments and computer simulations of structural changes in complex molecular systems have shown increased *order*, such as crystallization, under conditions when the freedom of movement increased. The First Law of Thermodynamics states that energy is conserved, but Nature has the freedom to distribute that energy. It does this in such a way that, according to the Second Law, the total freedom of movement is maximized. In some cases this may lead to disorder, but in other cases it is the opposite and the freedom of movement may increase considerably by ordering, by crystallization. Maximizing freedom of movement is a fundamental driving force in nature. And this force is not only valid in the world of physics, living matter too is subject to the Second Law of Thermodynamics and cells and organisms continuously strive to increase their freedom of movement. A study, recently published in the journal Science, indicates that double-stranded DNA is increasing its freedom of movement by curling up into the well-ordered double helix structure. Thus maximizing the freedom of movement is one of the fundamental driving forces not only among atoms and molecules, but also in the origin **and** the evolution of life.

Today there is a new separation of cultures. The science/literature polarity is overshadowed by the opposing interpretations of the Second Law of Thermodynamics. On the one hand a rather pessimistic worldview prevails, both in modern science as well as in literature, where it is believed that nature's fundamental driving force leads to maximum disorder and chaos. As we all know humans have been responsible for creating disorder and chaos, particularly in the twentieth century and with the help of scientists. However, the notion frequently expressed in literature that this is legitimated by and almost inevitable because of a fundamental driving force of nature is unfounded. It is a widely held misconception of the Second Law.

Fortunately there is another, a more positive view supported by recent theory and experiment; it says that nature requires energy in order to maximize its freedom of movement. It should make all the difference in our culture, science **and** literature, if, instead of viewing our world as driven towards disorder, its driving force, its arrow of time, is to increase the freedom of movement.

With this I hope to have cleared the looking glass somewhat. I will finish with one more example. Let us imagine this ceremony in the Pieterskerk. For the moment you are all seated in well-ordered rows but at the end of the ceremony you will leave your seats and move to the back of the church for the reception. At that stage, you may call our meeting disordered; on the other hand you could equally well say that our driving force is not to maximize disorder but to increase our freedom of movement, to meet with others, to open new opportunities, to start new relationships and developments, to progress. Our founding father had an inkling of this also, for you know the motto of Leiden University: *Praesidium Libertatis* (Stronghold of Freedom). Isn't this more optimistic worldview worth celebrating?

2005

With special thanks to my friends in the
"DISPUUT OPGERICHT ZATERDAG 16 NOVEMBER 1946"

Sir Charles

There is more than grandeur in your view of life. Indeed, natural selection has led to that grandiose biodiversity from single-celled to the most complex organism – humans – but today we know that your theory also explains social behaviour and culture, including human morality. Here I think it relevant to tell you about work done by three Dutchmen of whom you may have known one; I will save him for last.

First comes Frans de Waal, last year selected by TIME as one of the hundred most influential people in the world, because of his research on chimpanzees and others apes in Burgers Zoo and in Yerkes Field Station. In his most recent book, *Primates and Philosophers: How Morality Evolved*, he tells about two chimps performing the same task for which one receives a peanut, the other what they prefer much more: a grape. Soon it turns out that the one who gets peanuts does not think it fair and goes on strike. In another experiment two apes have to pull a heavy load to reach their food. Individually it does not work, but in good harmony they succeed. It is very funny to see how they manage to do it and when one ape is not hungry anymore, the other slaps him on the back encouraging him to continue, and this works. In a less friendly experiment two animals are put in cages next to each other and one gets an electric shock every time the other takes some food. The other apparently cannot bear this responsibility and refuses to eat. Frans de Waal concludes that these apes show the most important elements of moral behaviour: 1. Empathy and altruism; 2. Fairness and social cohesion. These experiments make clear that consciousness, ethics and morality are no longer the exclusive subjects of philosophers and theologians.

Sir Charles, in your *Descent of Man* you already anticipated this but now it has been shown experimentally: moral behaviour emerged in evolution long ago, long before humans did.

The second Dutchman about whom I want to tell you is Niko Tinbergen, the biologist, who in 1963 published a famous article, *On Aims and Methods of*

Ethology, in which he introduced the four why-questions of behavioural biology. The first question is: What is the immediate cause for certain behaviour? The second question: How does behaviour come about, is it inherited or learned? The third question concerns the evolution of behaviour in successive generations. The fourth and for Tinbergen by far the most important question is: What is the function of behaviour? What is its survival value? Answering this question earned him the Nobel Prize.

Sir Charles, you will agree with him that the question on the survival value of behaviour is the most important. In evolutionary thinking moral behaviour contributes to survival. The behaviour of humans and other animals shows many mutations, most are selected away, they disappear like snow in the sun, but that behaviour, that culture, which really contributes to our survival, that culture by its very nature will also survive. Some people think that morality was imposed on us by heaven through Moses' tablets; today we know that the origin of the Ten Commandments in our evolution is indeed by means of natural selection. In Darwinian thinking the value of morality lies in its contribution to the survival of the individual, of the family, of the species, of society, of life on earth.

The third Dutchman knew all this already and you may have known him. He was not a scientist, but a famous writer and his book, like your *Origin of Species*, also appeared in 1859. It has become the most famous book in Dutch literature and was translated into many languages. It is also the most moral book in our language. Perhaps that is why it is a classic? For us there is every reason to celebrate Mutatuli's *Max Havelaar* this year, as well as your *Origin of Species*.

Yours truly,
Frans W. Saris
2009

Darwin meets Einstein

Dialogues on the value of science

Frans W. Saris

This is a play in seven short scenes for two actors and two singers. At the start and at the end of the play, and twice in-between scenes, Leiden University students sing about love. After each scene one actor stays on stage while the other disappears and returns as a new character. Every scene takes place on the same set: the Leiden Botanical Gardens. Instead of the Leiden beech tree one sees the Tree of Life in various stages of its evolution.

First performance 31 August 2007, (in English) 15 December 2007 by
Jan Kijne: Kamerlingh Onnes, Albert Einstein, Spinoza, Francis Bacon:
Frans Saris: Francis Bacon, Franz Kafka, Charles Darwin, Niko Tinbergen:
Merlijn Runia: mezzo-soprano, *Chanson d'amour* (Fauré), *A Chloris* (Hahn),
    Beau soir (Debussy) and *La ci darem la mano* (Mozart)
Aleksei Nazarov: baritone, *Sred' shumnogo bala* (Tchaikovsky) and *La ci darem la mano* (Mozart)
Anke van der Kooy: piano
Richard Todd: translation

Darwin meets Einstein

Dialogues on the value of science

Students sing about love

Scene 1. Solomon's House

Francis Bacon:

Good evening. Could you please tell me where I can find the famous Leiden beech tree?

Kamerlingh Onnes:

Oh, it was so old; it has become one of us.

Francis Bacon:

Didn't the students propose to each other under this tree?

Kamerlingh Onnes:

It was also Lorentz's and Einstein's favourite place. Zeeman and I liked to meet here, Huizinga lived across the street. You could do business here, talk and think, this was the heart of the university.

Francis Bacon:

I don't think students propose to each other here anymore.

Kamerlingh Onnes:

What is more, the university has become a multiversity.

Francis Bacon:

All gone, along with the Leiden beech tree?

Kamerlingh Onnes:

I can't believe that. Students still like the Botanical Garden to make love. For some of our scientists this is still their favourite spot, they meet up here. The strange thing about the dead is, you only have to call them by name, you only have to think of them, and they present themselves at once. You could also say the Leiden beech tree is for the right people, the right place at the right time.

Francis Bacon:

Excuse me, may I introduce myself: Francis Bacon.

Kamerlingh Onnes:

Kamerlingh Onnes is my name. Yes, I know you from the *New Atlantis*.

Francis Bacon:

And *Solomon's House*.

Kamerlingh Onnes:

Yes, I've organized my laboratory according to its principles. With all its instruments and well-trained researchers and that whole army of blue-collar workers who helped me operate and service all the equipment. Your magnificent description of those who helped gather all knowledge, you called them ...

Francis Bacon:

'Merchants of light' ...

Kamerlingh Onnes:

And the clever engineers who could think up the most mysterious measurements, you called them ...

Francis Bacon:

'Mystery men' ...

Kamerlingh Onnes:

And then those who were completely free to do whatever experiment they wanted to do ...

Francis Bacon:

'Pioneers or miners'.

Kamerlingh Onnes:

The writers who recorded all the measurements ...

Francis Bacon:

'Compilers'.

Kamerlingh Onnes:

Those who interpreted the data and extracted the discoveries so that all our endeavours were at least worthwhile ...

Francis Bacon:

'Dowry-men or benefactors'.

Kamerlingh Onnes:

And let us not forget those who interpreted our experimental discoveries as higher observations, axioms and aphorisms, and I think that for you these were most important of all and you called them ...

Francis Bacon:

'Interpreters of nature'.

Kamerlingh Onnes:

My biographer believes that I discovered the system of 'bachelor-master-fellow' in that orchestrated collaboration which today we call 'Big Science'. The way modern science should be done, however, you (Sir Francis) already described it in the seventeenth century ...

Francis Bacon:

Solomon's House had four aspects: a role for all the collaborators high and low; a requirement for every piece of equipment and all the instruments; and the importance of rituals and ceremonies ...

Kamerlingh Onnes:

Yes, sometimes it seems as though science is a competition, with team leaders, coaches, suppliers, sponsors, and supporters with only one goal in mind: the Nobel Prize.

Francis Bacon:

Is science not concerned first and foremost with the advancement of knowledge? Is that not the fourth and most important aspect of science?

Kamerlingh Onnes:

No, not knowledge for its own sake – or indeed art for art's sake.

Francis Bacon:

Were you not awarded the Nobel Prize for the liquefaction of helium? And how did that benefit mankind?

Kamerlingh Onnes:

Most people also think I was engaged in a competition, a race towards 'absolute zero', however, as my biographer correctly observed: 'In the latter part of the nineteenth century electrotechnics conquered the world; in the beginning of the twentieth century that role was arrogated to the business of chilling. So we had beer brewers and ice-cream manufacturers, cold stores and cold wagons, hospitals, milk, chocolate, gum, and perfume factories, the textile industry and manufacturers of liquid carbonic acid, ammonia and air.' In short, cold was both a hype and a contribution to an international growth market.

Francis Bacon:

So, science became a matter of creating wealth, rather than what I thought it was, something that worked towards our well-being.

Scene 2. Before the Law

Franz Kafka:

Good day to you. Franz Kafka is the name. Are you not Professor Kamerlingh Onnes, known for: 'To measure is to know'? In 1914 – you had just been awarded the Nobel Prize for super-conductivity – I wrote a short parable.

Kamerlingh Onnes:

The Nobel Prize was for my work on liquid helium, not for super-conductivity. We did measure endless ring-currents, so we know superconductivity exists, but we do not have any explanation for it yet. We are faced with great enigmas.

Frans Kafka:

Speaking of enigmas, may I read my short parable to you?

From *The Trial*, RT based on trans
© David Wyllie (Project Gutenberg EBook)

Before the Law there is a doorkeeper. A man from the countryside comes up to the door and asks for entry. But the doorkeeper says he cannot let him in to the Law at this moment. The man thinks about this, and then he asks if he will be able to go in later on. 'Possibly,' says the doorkeeper, 'but not now.' The gateway to the Law is open as it always is, and the doorkeeper has stepped to one side, so the man bends over to try and look in. When the doorkeeper notices this he laughs and says, 'If you're tempted give it a try, try and go in even though I say you can't. Take care, though: I am powerful. And I'm only the lowliest of all the doormen. But there's a doorkeeper for each of the rooms and each of them is more powerful than the last. It's more than I can stand just to look at the third one.' The man from the countryside had not expected difficulties like this: the Law was supposed to be accessible for anyone at any time, he thinks, but now he looks more closely at the doorkeeper in his fur coat, sees his big hooked nose, his long thin tartar-beard, and he decides it's better to wait until he has permission to enter. The doorkeeper gives him a stool and lets him sit down to one side of the gate. He sits there for days and years. He tries to be allowed in time and again and tires the doorkeeper with his requests. The doorkeeper often questions him, asking about where he is from and many other things, but these are disinterested questions such as great men ask, and he always ends up by telling him he still cannot let him in. The man had come well equipped for his journey, and uses everything, however valuable, to bribe the doorkeeper. He accepts everything, but as he does so he says, 'I'll only accept this so that you don't think there's anything you've failed to do.' Over many years, the man watches the doorkeeper almost without a break. He forgets about the other doormen, and begins to think this one is the only thing stopping him from gaining access to the Law. Over the first few years he curses his unhappy condition out loud, but later, as he becomes old, he just grumbles to himself. He becomes senile, and as he has come to know even the fleas in the doorkeeper's fur collar over the years that he has been studying him. He even asks *them* to help him and change the doorkeeper's mind. Finally his eyes grow dim, and he no longer knows whether it is really becoming crepuscular or whether it is his eyes that are deceiving him. But now he seems to see an immortal radiance beginning to shine from the darkness behind the door. He does not have long to live now. Just before he dies, he gathers all his experience, from all the time he has waited, into one question which he has never yet put to the doorkeeper. He beckons to the doorkeeper, as he is no longer able to raise his stiff body. The doorkeeper has to bend over deeply as the difference in their sizes has changed greatly to the disadvantage of the man from the countryside. 'What do you want to know now?' asks the doorkeeper, 'You're absolutely insatiable.' 'Everyone wants access to the Law,' says the man, 'how come, over all these years, no-one but me has ever asked to be let in?' The doorkeeper can see the man from the countryside has come to his end, his hearing has faded, and so, in order that he can be heard, he shouts to him: 'Nobody else could have got in this way, because this entrance was meant only for you. Now I will go and close it.'

Halfway through the parable Kamerlingh Onnes shrugs his shoulders and leaves. Albert Einstein enters and listens.

Scene 3. Laws of Nature

Albert Einstein:
   So the doorkeeper deluded the man.
Franz Kafka:
   Professor Einstein, did *you* read my parable?
Albert Einstein:
   Dear Kafka, of course I know that strange parable of yours. When I read it first, many years ago, I was still convinced that the whole of nature could be understood by means of physics and described by means of mathematics. Mathematics is the language of physics, that is true, but it is with man just as with the man from the countryside in your story: we can go as far as the door to the Law – the law of nature – we should very much like to enter and understand and know what is behind the mathematics, but alas, that is not given to us.
Franz Kafka:
   So, you are lost in Solomon's House!
Albert Einstein:
   Some people think that the key to Franz Kafka is your Letter to your Father; that your strange parables are figures for the poor relationship you had with your father. That is Freud of course, but please do recognize the strangeness of physics in the strangeness of your stories. Ever since the Enlightenment, we have been tempted to enter Solomon's House, but in front of one of the gates stands Francis Bacon and it is as if he bellows in our ears ...

Franz Kafka:

'This entrance was meant only for you. Now I will go and close it.' Do you feel he deluded you?

Albert Einstein:

You tell me, dear Kafka, what little is left of our mechanical world view of the nineteenth century? The physics of the twentieth century has become strange, even to me.

Franz Kafka:

The world is a nightmare. Your relativity theory, quantum mechanics, the standard model, the cosmology, string theory. They have all been discovered in the twentieth century, but our world view remains completely obscure.

Albert Einstein:

Laws, yes – but what those laws really mean, what is behind those mathematical formulas, why they are like that, that is what we still do not understand. Take for instance the electron, the elementary particle we have known about for over a century. Its behaviour is described by quantum electrodynamics, but is the electron a wave or a particle?

Franz Kafka:

Yet we live in modern micro-electronic times. Today physicists send electrons in all directions: criss-cross into the grid, radio waves, TV channels, DVD players, in computers, mobile phones, and the world wide web, but still they do not know what the electron actually is or what it is made of. In fact, you wonder, do they actually know what they are doing? Isn't that frightening?

Albert Einstein:

Just as with the electron, so it is with the particle of light, the photon. We can do all sorts of things with them but we still do not really know what a photon is. In one experiment it behaves like a particle, and in another like a wave, do you understand that? The electron has no extension but it does have mass, who understands that? And it also has spin: it rotates about its own axis whereas it does not have any extension, how on earth is that possible?

Franz Kafka:

$E = mc^2$ – your own formula – everyone knows, we even wear it on our T-shirts, but we still don't understand it.

Albert Einstein:

Yes, mass and energy are equivalent, and with this principle we have built the atom bomb – but where mass comes from, how gravity can be brought into line with the other laws of nature – that we still don't know.

We live in a universe that is curved in space and time, the experiments clearly prove it, but can you envisage how?

Franz Kafka:

Yes, I can. It is a labyrinth. But what to me really is an enigma, what I cannot possibly comprehend, is the Big Bang and the idea that our universe has been expanding ever since then. Into what is it expanding? Into a vacuum? Into nothing at all?

Albert Einstein:

My dear Kafka, for me that is also a very sensitive issue and it has not become any clearer yet. Astronomers claim to have a clear view of the origin and evolution of our universe, but it has now become evident that they have only detected 5% of all matter and radiation, 95% remains invisible. They call it dark matter and dark energy, it has to exist, perhaps even here and now, but we cannot see it. Can you imagine: we conceive a picture of the universe, whereas most of it is as yet impossible to imagine?

Franz Kafka:

And that enigma is not resolved by the so-called string theory.

Albert Einstein:

Oh no, because we will never be able to test that string theory experimentally, its dimensions are much too small for that. Again, it is one of those Laws we will not be allowed to gain access to.

Franz Kafka:

It would be much more mischievous if I were to ask you: what is the *electron* made of? You would answer: it is made of strings. Then my next question would be: what are strings made of? So the string theory cannot possibly hold the long- awaited Theory of Everything.

Albert Einstein:

Oh, that Theory of Everything! It keeps bothering me, it is a true obsession. I have spent a considerable part of my life on it, in vain. We can do almost anything in this world; our society has been dramatically changed by science and technology. But what is the value of science if we cannot find the Theory of Everything? I am in great despair. Indeed I feel like the man from the countryside who wanted to enter the Law, but at the end of his life the doorkeeper bellowed in his deaf ears:

Franz Kafka:

'Nobody else could have got in this way, as this entrance was meant only for you. Now I will go and close it.'

Both leave shaking their heads

FRANS W. SARIS

Student sings lovesong

Scene 4. Theory of Everything

Charles Darwin:
   My dear Einstein, what are you so worried about? Still longing for the
   Theory of Everything? Whereas it already exists.
Albert Einstein:
   What? Have I missed something?
Charles Darwin:
   Yes, I believe you have, but you are by no means alone.
Albert Einstein:
   Do you mean there is a Theory of Everything, but that until now it has
   been overlooked?
Charles Darwin:
   Yes, and many people even rejected this theory as heresy.
Albert Einstein:
   I don't understand: please explain.
Charles Darwin:
   Well, first there are the physicists, like you, the hardcore reductionists,
   who think biology is nothing but chemistry and chemistry is nothing but
   physics and therefore the Theory of Everything should be found in ele-
   mentary particle physics.
Albert Einstein:
   What is wrong with that?
Charles Darwin:
   Science does not work that way, you will never get anywhere along that
   road. If you had been able to, there would otherwise surely have been a
   Theory of Everything a long time ago?
Albert Einstein:
   But you say there already is a Theory of Everything?
Charles Darwin:
   Yes, but not as far as those arrogant physicists are concerned. Everything
   is physics, but physics is not everything. In addition, as I said, there are
   many people who consider my Theory of Everything a heresy.
Albert Einstein:
   Who are you anyway?
Charles Darwin:
   The fact that you don't know is revealing, but I don't mind. My name is
   Charles Darwin.

Albert Einstein:

But of course, I apologize for not recognizing you! Charles Darwin, and the Theory of Evolution.

Charles Darwin:

Precisely. *The Origin of Species, by means of natural selection* dates from 1859 and was meant as a theory for natural history, as biology was called back then, but in the meantime the theory of natural selection seems also to have proved applicable to many other fields of scientific endeavour, perhaps even all sciences and maybe every development in our world. Some people regard the theory of evolution fruitful not only in the natural sciences but also in many other aspects throughout our culture.

Albert Einstein:

Including ethics and moral behaviour?

Charles Darwin:

Yes, ethics and moral behaviour too.

Albert Einstein:

Music, literature, and visual arts?

Charles Darwin:

Some people think they are by-products, which evolved by accident and do not have any function, but I cannot believe that.

Albert Einstein:

Then the Theory of Evolution might also explain medicine, economy, politics, and even religion?

Charles Darwin:

Yes, although I am not yet convinced of an evolutionary explanation of religion.

Albert Einstein:

So you truly postulate a Theory of Everything. That is almost unbelievable.

Charles Darwin:

I agree and I did not invent this all by myself, but the more I think about it, the more I become convinced that everything in our world evolves according to a theory of natural selection.

Albert Einstein:

What do you mean?

Charles Darwin:

You know all the elements of the Theory of Evolution: mutations – selection – reproduction. This holds not only for biology, not only for the evolution of the natural world and the Tree of Life, but also for our culture. Our culture, too, evolves according to those three elements: mutations, selection and reproduction of information to following generations, but cultural evolution proceeds much, much faster than biological evolution.

Albert Einstein:

What is the criterion for cultural selection? It seems to me to be crucial to know how selection comes about.

Charles Darwin:

Survival is the answer! I know it sounds tautological but survival is the only thing that counts.

Albert Einstein:

You mean that is true for both biological and cultural evolution?

Charles Darwin:

The more I think about it, the more I start to believe it really is so. The very mutation, in biology and in culture, which contributes to our survival, the survival of the individual, the family or the species – that mutation itself will also survive.

Albert Einstein:

Fantastic, unbelievable! Why did I not think about this earlier? Your theory is not concerned with what everything is made of, but with how everything works! I very much regret that I have not encountered you under the Leiden beech tree before, and I am deeply grateful for your insight. This I must tell Franz Kafka. So this Theory of Everything belongs in the curriculum of all our students.

Scene 5. God

Charles Darwin:

My dear Spinoza, it is my good fortune to see you here. I came to Leiden especially to meet with you.

Spinoza:

Charles Darwin, my pleasure, at your service.

Charles Darwin:

Are you familiar with the Theory of Evolution?

Spinoza:

But of course.

Charles Darwin:

Do you know why I waited for such a long time before publishing *The Origin of Species*?

Spinoza:

Yes, I know – and perhaps you know that I postponed publishing my *Ethics* even longer than you did; indeed, it appeared only after my death, and my delay was for precisely the same reasons as yours.

Charles Darwin:

It's because of God. I could not deny my faith in God and certainly not publicly.

Spinoza:

That is right, I realized my work would be considered heretical and I would be known as an apostate – whereas God or Nature, who for me are the same, I admire wholeheartedly.

Charles Darwin:

I was told that here at Leiden University (and not only here), the same question is still being asked, and not only by philosophers. It is a rather rhetorical question: in what kind of God do people believe who believe both in God and in the Theory of Evolution?

Spinoza:

That question is certainly not meant as purely rhetorical; people ask it nowadays who as atheists fight against religion almost as fundamentalists do.

Charles Darwin:

Whereas there is a very different question, in respect of religion, that I consider much more fundamental.

Spinoza:

Do you mean: is there also an evolutionary theory of religion?

Charles Darwin:

Exactly!

Spinoza:

Well, if you pose the question you must also try to answer it.

Charles Darwin:

I don't think the first hominids knew religion. Apes are not religious, are they?

Spinoza:

Somewhere there must have been a people who discovered religion for the very first time.

Charles Darwin:

When people started to practice agriculture, instead of hunting and gathering, they also experienced unanticipated problems. After harvesting, the temptation to overindulge, to overeat and drink too much, was irresistible, so that frequently all the supplies that had been gathered were quickly exhausted. People did not consider storing anything, and as a result there was nothing left for a bad day.

Spinoza:

They therefore came to feel the need for rituals, and looked to priests for guidance and for taking care of the future. In this way religion may have emerged.

Charles Darwin:

People who chose religion were much better prepared for the disastrous times. During such times they found more help and consolation because there was quite simply much more social cohesion and thus a much larger probability of survival than for people lacking a religion. As a result religion spread all over the world and indeed today we only find peoples who engage in whatever way with some form of religion.

Spinoza:

The conclusion must be that religion has given evolutionary advantage to people. Is that still the case today?

Charles Darwin:

This I think is where the shoe pinches. Churches have misused their powerful position to such an extent that because of religious fundamentalism and terrorism, religion has played into atheists' hands.

Spinoza:

Moreover, since the Enlightenment, most people do not believe in the literal truth of the Bible, just as they don't believe in a personal God who providentially directs their personal life.

Charles Darwin:

This is potentially dangerous, because if religion really does contribute to survival, how can people do without it?

Spinoza:

No, I suppose they can't.

Charles Darwin:

That is why we still have to answer the question: in what kind of God do people believe who believe in God and in the Theory of Evolution?

Spinoza:

It appears to be a difficult question.

Charles Darwin:

A satisfactory answer seems almost impossible. That is why I waited such a long time before publishing *The Origin of Species*.

Spinoza:

Yet the answer to your question is very simple, even immaterial: God and Nature, whatever you want to call them, are one and the same.

Charles Darwin:

God is Nature, yes, but then how about religious ritual and the so-called

Holy Scriptures, revelation, and moral behaviour, can it simply be abandoned?

Spinoza:

We may believe in Nature as we may believe in God, while we may also believe in evolution. In addition there is the matter of ethical behaviour: that is, the ethics of survival. And God or Nature also creates a close companionship with nature, God in us. In due course I have learned to read the Bible purely as metaphor and to see religious ritual in a similar manner. During the course of our cultural evolution these things have evolved too and have remained with us, so they clearly perform a function. They have survival value.

Student sings lovesong

Scene 6. Evolution of Science

Niko Tinbergen:

Dear Spinoza, I should like to discuss with you the evolution of science.

Spinoza:

My dear Niko Tinbergen, I wonder whether by the evolution of science you mean that first there were people lacking any sense of science, then during the Age of Enlightenment modern science was invented by Francis Bacon. Peoples armed with modern science enjoyed a larger probability of survival than peoples without, so that nowadays everyone strives towards modern science and technology.

Niko Tinbergen:

Indeed, that is what I mean, but I would like to take it a little further. As an ethologist I have studied the behaviour of animals, always asking the same four questions. I like to pose those questions to people who practice scientific research as well. My first question is: what is the immediate reason that humans carry out scientific research?

Spinoza:

That is very different for different people. There are those who do it out of competitiveness: they want to be first in their field, and for them it is only the first prize that counts. Others are more like pioneers; they want to be first to reach a peak or an uninhabited area. Yet others are much more motivated in improving our world, they practice science for the benefit of mankind.

Niko Tinbergen:

My second question is: how does inquiry, how does the behaviour of scientists, come about, is it taught or inherited?

Spinoza:

Both surely. People are born as curious animals, and for some of them science is a passion, so that in principle education should stimulate inquiry. Curiosity is important, if you do not look you won't find anything at all, but if on the other hand you *are* looking for something you will frequently find something very different from what you had expected. Once you experience the thrill of the creative moment, you become intoxicated, even hooked.

Niko Tinbergen:

And now here is my third question: how did scientists *develop*? How did the practice of science evolve over the course of time?

Spinoza:

You know that better than I do. In my days scientists still worked individually: today they perform their work in orchestrated groups from very different laboratories all over the world, all of whom try to solve the same given problem collaboratively. 'Big Science for Big Business' is not only true in physics. Even life scientists have to work in this way and in due course the social sciences and even the humanities will have to follow suit.

Niko Tinbergen:

My fourth and by far most important question is this: what is the function of science?

Spinoza:

I agree that this is indeed the most compelling question, and in my lifetime and that of Sir Francis Bacon the answer was very clear. Today scientists have lost track of what they are supposed to be doing. Science has become a contest, and a hype, as a result a certain decadence has set in. No wonder society at large does not value scientific research and university education as much as it should.

Niko Tinbergen:

In evolutionary thinking, science contributes to our survival. That is its function. Individual behaviour within our culture displays many mutations. Most mutations are selected out: they melt like snow in the sun, except for those behaviours, and that culture, which truly contributes to our survival. That culture will certainly survive. This holds true especially for the sciences. The value of science is, with God or Nature as our conscience, to contribute to our survival, to the survival of the individual, the family, the species, life on earth. It is this that makes Natural and Human Sciences worthwhile.

Spinoza:

Excellent! We must try to convince our students.

Niko Tinbergen:

No, it is the professors we must convince! I still remember how perplexed I was upon being told off firmly by one of my Zoology professors when I raised the idea of the value of survival after he had asked: 'Has anyone any idea why so many birds flock more densely when they are attacked by a bird of prey?'

Scene 7. The Value of Science

Francis Bacon:

Solomon's House has acquired a solid position in society. Undoubtedly, the Enlightenment was the proper time for an industrial revolution, for technological development and economic growth, for globalization, for improvement of our diet, hygiene and medical care. Progress, wealth and well-being, all thanks to science.

Niko Tinbergen:

Sir Francis, science cannot be wished out of this world, but is it still moving in the right direction? Today there is hardly any belief in progress. During the industrial revolution, a time of economic growth coexisted with a substantial reduction in life expectancy among the workers; voyages of discovery coexisted with colonization and slavery; technological development coexisted with dreadful wars and genocides; improved diet, hygiene and medical care coexisted with a dramatic population explosion; world trade coexisted with exploitation and environmental damage; modern energy supplies coexisted with smog and climate change; bio-industry coexisted with a reduction in biodiversity. And the scientists have gone into hiding in what they call science devoid of value.

Francis Bacon:

But surely science, since the Age of Enlightenment, is what we ought to be most proud of.

Niko Tinbergen:

Yes, I agree, but the more we know the less we believe. Our world view has become nihilistic, aimless, without any moral initiative. Before Copernicus, the earth was still the centre of our universe, before Darwin people were created in the image of God. Through astronomy we now know how infinitely large and empty our universe is, continually expanding since the Big Bang. Because of the theory of evolution we have come to understand the place of humankind on the tree of life – as one tiny twig that could have been as insignificant as countless others, thanks to a magnificent accident. If one could rewind the film of evolution all the way to the beginning and replay it, would we reappear in a form that we would recognize? Probably not. Why then are we here? If there is neither God nor commandment, then how are we supposed to live?

Francis Bacon:

You are forgetting Spinoza with his Nature-as-God! And Darwin with his Theory of Everything! And in your own field, behavioural biology, Frans de Waal was chosen by TIME magazine as one of the hundred most influential people in the world. His research during the past decades has shown that chimpanzees and other apes possess most of the characteristics of moral behaviour: first empathy and altruism, secondly reciprocity and fairness. With that our sense of conscience, ethics and moral behaviour are no longer exclusively the preserve of theologians and philosophers, but concern biologists even more. Moral behaviour evolved long before humans did. Its function is absolutely clear: survival. Frans de Waal's research has liberated us from original sin. Humans are not necessarily programmed to commit evil!

Niko Tinbergen:

At the same time scholarly knowledge has become fragmented into all sorts of specialisms, not only in science and humanities, but at the same time tribal wars are raging within each discipline, tribal wars for life or death, that is to say for the money. We no longer appreciate each other's discourse or culture, neither does *homo universalis* exist anymore. No wonder scientists have lost track and have converted their endeavours into a contest, valued only in terms of numbers of top publications, citations, prizes and subsidies, hypes irrespective of any goal. Large numbers of scientists contributed to the Cold War and to the development of WMDs; others unscrupulously modify genetic material of plants, animals and humans without knowing exactly what the consequences might be. The best science of all must be fundamental, which is the equivalent of useless. Isn't that the depth of decadence? Although our society has become completely dependent on science and technology, and our world view has been dramatically altered by scientific discoveries, the position of science in our postmodern culture has been marginalized. And what happens to the idealism with which young students enter the Academy, expecting to be able to contribute to a better world, if the Academy does not care?

Francis Bacon:

Science not only improved the quality of life for many people; it's also true that because of science we know that economic developments in the Western world are now out of control.

Niko Tinbergen:

Yes, the footprint of the Western world is disproportionately large, larger than the carrying capacity of our globe. If we continue in this way we are heading for a global catastrophe. Economic growth is not sustainable, certainly if China and India join the global economy, and what about Africa? The way our society is organized will have to change in a revolutionary way, and if it does not we may expect gigantic instabilities, mass migrations, terrorism, world wars, natural disasters through climate change, and perhaps even mother earth becoming uninhabitable for us humans.

Francis Bacon:

But humans are the first species able to create the circumstances in which future generations will have to live, no species before us could do this and thus we are the first to influence our own evolution. As our culture evolved, so also our conscience developed into a pyramid of morality. We feel responsible not only for our own lives and that of our kin, but also for the survival of our species. Today we may be able to detect on the horizon a glimpse of responsibility for life on our planet. Today we hear not only about personal and national conscience, but also about global conscience.

Niko Tinbergen:

Today the door to Solomon's House is not closed. Sustainable development is needed, not only in the fields of energy, resources, and industrial development, also in the area of water management, biodiversity, nature conservation and production of food. Globalization has to be sustainable, which means taking into account diversity in culture, politics, and religion. This is the challenge for all scientists, scholars, students, masters and fellows, and in all disciplines: alpha, beta and gamma. This is what makes natural and human sciences worthwhile.

Francis Bacon:

Indeed, the door to Solomon's House is widely open to those scientists, scholars and students who are not inward-looking but open to the outside world.

Students sing lovesong

© Frans W. Saris and Richard Todd
14 December 2007

To colleagues and friends, for decades of Fun,
Utilization, Theories of everything and Survival,
Thank you

For Product Safety Concerns and Information please contact our EU
representative GPSR@taylorandfrancis.com
Taylor & Francis Verlag GmbH, Kaufingerstraße 24, 80331 München, Germany